Getting Started with CockroachDB

A guide to using a modern, cloud-native, and distributed SQL database for your data-intensive apps

Kishen Das Kondabagilu Rajanna

BIRMINGHAM—MUMBAI

Getting Started with CockroachDB

Publishing Product Manager: Sunith Shetty
Senior Editors: Roshan Kumar, Nazia Shaikh
Content Development Editor: Tazeen Shaikh
Technical Editor: Rahul Limbachiya
Copy Editor: Safis Editing
Project Coordinator: Aparna Ravikumar Nair
Proofreader: Safis Editing
Indexer: Pratik Shirodkar
Production Designer: Jyoti Chauhan
Marketing Coordinator: Priyanka Mhatre

First published: March 2022

Production reference: 1090322

Published by Packt Publishing Ltd.
Livery Place
35 Livery Street
Birmingham
B3 2PB, UK.

ISBN 978-1-80056-065-9

www.packt.com

During my childhood, I have spent more time watching insects than interacting with human beings. So, I dedicate this book to cockroaches and all the other insects that have fascinated me throughout my life.

Contributors

About the author

Kishen Das Kondabagilu Rajanna is currently working as a distributed query engineer at Adobe. His previous experience includes leading the data warehouse team at Cloudera, managing the SaaS platform at Rubrik, writing core services for the Oracle public cloud, and managing data infrastructure at Apple.

About the reviewers

Nadir Doctor is a database and data warehousing architect, plus DBA, who has worked in various industries with multiple OLTP and OLAP technologies, as well as primary data platforms, including CockroachDB, Snowflake, Databricks, DataStax, Cassandra, ScyllaDB, Redis, MS SQL Server, Oracle, Db2 Cloud, AWS, Azure, and GCP. A major focus of his is health check scripting for security, high availability, performance optimization, cost reduction, and operational excellence. He has presented at several technical conference events, is active in user group participation, and can be reached on LinkedIn.

Thank you to Kishen and all the staff at Packt. I'm grateful for the immense support of my loving wife, children, and family during the technical review of this book. I hope that you all find the content enjoyable, inspiring, and useful.

Scott Ling is a technology specialist with over 30 years experience of working at the forefront of technology in various roles, in companies from start-ups to 10 bn+ listed companies, with a focus on distributed technologies, software as a service, and product/project management. He is currently working on a free product designed to make it easy for anyone to create and manage an SaaS product, service, or business as his way of giving back to the community. Scott is also an established technical author with a bestselling book on .NET published back in 2001 and has worked with authors and publishers on various books and technologies over the years.

Table of Contents

Preface

Section 1: Getting to Know CockroachDB

1

CockroachDB – A Brief Introduction

The history and evolution of databases	3	CAP theorem	13
SQL	4	Consistency and partition tolerance (CP)	13
Object-oriented databases	4	Availability and partition tolerance (AP)	14
NoSQL	5	Consistency and availability (CA)	15
NewSQL	5	CockroachDB	15
Database concepts	6	Why yet another database?	16
Cardinality	6	Inspiration	16
Overview of database models	8	Key terms and concepts	16
Processing models	11	High-level overview	17
Embedded and mobile databases	12	Summary	18
Database storage engines	12		

2

How Does CockroachDB Work Internally?

Technical requirements	20	Parsing	25
Installing a single-node CockroachDB cluster using Docker	21	Logical planning	25
		Physical planning	26
		Query execution	30
Execution of a SQL query	24	Managing a transactional key-value store	31
SQL query execution	24		

Data distribution across multiple nodes 32

The MSKVS 32
Meta ranges 32
Table data 34

Data replication for resilience and availability 34

What is consensus? 34
The Raft distributed consensus protocol 34

Interactions with the disk for data storage 41

Storage engine 42

Summary 43

Section 2: Exploring the Important Features of CockroachDB

3

Atomicity, Consistency, Isolation, and Durability (ACID)

An overview of ACID properties 48
Atomicity 48
Consistency 49
Isolation 49
Durability 51

ACID from CockroachDB's perspective 51
Atomicity 51
Consistency 57
Isolation 57
Durability 57

Summary 60

4

Geo-Partitioning

Technical requirements 62
Introduction to geo-partitioning 62
Cloud, regions, and zones 64
Region 65
Zone 65

Regions and zones on various cloud providers 66

Geo-partitioning in CockroachDB 67
Single region 67
Multi-region 70

Summary 84

5

Fault Tolerance and Auto-Rebalancing

Technical requirements	86	Working example of fault tolerance at play	87
Achieving fault tolerance	86		
Achieving fault tolerance at the storage layer	86	Automatic rebalancing	96
		Recovering from multi-node failures	98
		Summary	99

6

How Indexes Work in CockroachDB

Technical requirements	101	Inverted indexes	110
Introduction to indexes	102	Partial indexes	111
Different types of indexes	103	Spatial indexes	113
Primary indexes	104	Table joins and indexes	115
Secondary indexes	106	Best practices while using indexes	118
Hash-sharded indexes	108		
Duplicate indexes	110	Summary	120

Section 3: Working with CockroachDB

7

Schema Creation and Management

Technical requirements	124	Supported data types	136
DDL	124	Column-level constraints	137
CREATE	124	Table joins	140
ALTER	129	Using sequences	141
DROP	130	Managing schema changes	142
DML	131	Summary	143
DQL	135		

8

Exploring the Admin User Interface

Technical requirements	146	Understanding sessions	160
Introducing the admin UI	146	Transactions	162
Cluster overview	148	Tracking jobs	163
Metrics deep dive	151	Summary	165
Database and table definitions	158		

9

An Overview Of Security Aspects

Technical requirements	168	Data encryption at rest and in flight	177
Introduction to security concepts	168	Encryption at rest	177
Client and node authentication	169	Encryption in flight	178
Generating certificates and keys	171	Audit logging	178
Client authentication	172	RTO and RPO	181
Node authentication	173	Keeping the network secure	182
Authorization mechanisms	174	Security best practices	183
Roles	175	Summary	184
Privileges	176		

10

Troubleshooting Issues

Technical requirements	186	Tracking slow queries	191
Collecting debug logs	186	Capacity planning	192
Log files	186	Configuration issues	193
Log levels	187	Guidelines to avoid issues during an upgrade	193
Log channels	188		
Emitting logs to an external sink	189	Network latency	194
Gathering Cockroach debug logs	189	Advanced debugging options	196
Connection issues	190	Summary	197

11
Performance Benchmarking and Migration

Technical requirements	200	Migration – Things to consider	207
Performance – Things to consider	200	Migrating from traditional databases	208
Infrastructure	200	Migrating from PostgreSQL to CockroachDB	209
Popular benchmark suites	201		
Benchmarking your specific use cases	202	Summary	210
Performance benchmarking for CockroachDB	204		

Appendix:
Bibliography and Additional Resources

Index

Other Books You May Enjoy

Preface

This book will introduce you to the inner workings of CockroachDB and help you understand how it provides faster access to distributed data through a SQL interface. You'll learn how you can use the database to provide solutions that require data to be highly available.

Starting with CockroachDB's installation, setup, and configuration, this book will familiarize you with the database architecture and database design principles. You'll then discover several options that CockroachDB provides to store multiple copies of your data to ensure fast data access. The book covers the internals of CockroachDB, how to deploy and manage it on the cloud, performance tuning to get the best out of CockroachDB, and how to scale data across continents and serve it locally. In addition to this, you'll get to grips with fault tolerance and auto-rebalancing, how indexes work, and the CockroachDB Admin UI. The book will guide you in building scalable cloud services on top of CockroachDB, covering administrative and security aspects and tips for troubleshooting, performance enhancements, and a brief guideline on migrating from traditional databases.

By the end of this book, you'll have enough knowledge to manage your data on CockroachDB and interact with it from your application layer.

Who this book is for

Software engineers, database developers, database administrators, and anyone who wishes to learn about the features of CockroachDB and how to build database solutions that are fast, highly available, and cater to business-critical applications, will find this book useful. Although no prior exposure to CockroachDB is required, familiarity with database concepts will help you to get the most out of this book.

What this book covers

Chapter 1, CockroachDB – A Brief Introduction, talks about databases and how they have evolved over time. You will also get to know about the high-level architecture of CockroachDB.

Chapter 2, How Does CockroachDB Work Internally?, explores various layers of CockroachDB and some of its inner workings.

Chapter 3, Atomicity, Consistency, Isolation, and Durability (ACID), introduces you to ACID properties and how they are implemented in CockroachDB.

Chapter 4, Geo-Partitioning, explains the concept of geo-partitioning, why we need it, and what are the various options for geographically distributing the data using CockroachDB

Chapter 5, Fault Tolerance and Auto-Rebalancing, explores the concept of fault tolerance and auto-recovery strategies. It also covers a few experiments to understand these concepts better.

Chapter 6, How Indexes Work in CockroachDB, is all about database indexes, how they are useful in improving query performance, the different types of indexes that are supported in CockroachDB, and some of the best practices that you can follow while using indexes.

Chapter 7, Schema Creation and Management, introduces you to SQL syntaxes for DDL, DML, and DQL with examples, different data types available in CockroachDB, sequences, and how to manage schema changes.

Chapter 8, Exploring the Admin User Interface, explores the admin user interface that comes by default when you deploy a CockroachDB cluster. We examine all the metrics and other information that are available in the user interface and how they are useful in troubleshooting issues.

Chapter 9, An Overview of Security Aspects, touches upon the key security aspects that you have to pay attention to when using CockroachDB. We also learn about authentication, authorization, how to protect data with encryption while at rest and in flight, and how to achieve desired data protection by defining the correct strategies for RTO and RPO. Network security and some security best practices are also covered.

Chapter 10, Troubleshooting Issues, helps you in getting yourself familiarized with troubleshooting issues by collecting logs, looking at some of the metrics, understanding and tracking slow queries, and integrating logs with external sinks. This chapter also covers advanced debugging options at the end.

Chapter 11, Performance Benchmarking and Migration, discusses performance, what the key indicators of performance are, and how to measure them. You will also learn about migrating from other traditional databases to CockroachDB.

Appendix: Bibliography and Additional Resources, provides additional resource material that you can go through to become more familiar with CockroachDB.

To get the most out of this book

You should have access to the internet so that you can download CockroachDB and try it on your laptop.

Software/hardware covered in the book	OS requirements
4 GB RAM, 250 GB HDD or SSD	Windows, macOS, or Linux

If you are using the digital version of this book, we advise you to type the code yourself or access the code from the book's GitHub repository (a link is available in the next section). Doing so will help you avoid any potential errors related to the copying and pasting of code.

Download the example code files

You can download the example code files for this book from GitHub at `https://github.com/PacktPublishing/Getting-Started-with-CockroachDB`. In case there's an update to the code, it will be updated on the existing GitHub repository.

We also have other code bundles from our rich catalog of books and videos available at `https://github.com/PacktPublishing/`. Check them out!

Download the color images

We also provide a PDF file that has color images of the screenshots/diagrams used in this book. You can download it here: `https://static.packt-cdn.com/downloads/9781800560659_ColorImages.pdf`.

Conventions used

There are a number of text conventions used throughout this book.

`Code in text`: Indicates code words in the text, database table names, folder names, filenames, file extensions, pathnames, dummy URLs, user input, and Twitter handles. Here is an example: "Here, in the `SELECT` query, you should use `AS OF SYSTEM TIME follower_read_timestamp()`."

A block of code is set as follows:

```
SHOW TABLES
```

When we wish to draw your attention to a particular part of a code block, the relevant lines or items are set in bold:

```
DROP DATABASE <DATABASE_NAME>
DROP ROLE <ROLE_NAME>
DROP TABLE <TABLE_NAME>
```

Any command-line input or output is written as follows:

```
$ cockroach cert create-client <user_name> \
--certs-dir=<certs_directory> \
--ca-key=<CA_key_directory>
```

Bold: Indicates a new term, an important word, or words that you see on screen. For example, words in menus or dialog boxes appear in the text like this. Here is an example: "The **Sessions** dashboard gives information about all the active client sessions within the CockroachDB cluster."

> **Tips or Important Notes**
> Appear like this.

Get in touch

Feedback from our readers is always welcome.

General feedback: If you have questions about any aspect of this book, mention the book title in the subject of your message and email us at customercare@packtpub.com.

Errata: Although we have taken every care to ensure the accuracy of our content, mistakes do happen. If you have found a mistake in this book, we would be grateful if you would report this to us. Please visit www.packtpub.com/support/errata, selecting your book, clicking on the Errata Submission Form link, and entering the details.

Piracy: If you come across any illegal copies of our works in any form on the internet, we would be grateful if you would provide us with the location address or website name. Please contact us at copyright@packt.com with a link to the material.

If you are interested in becoming an author: If there is a topic that you have expertise in and you are interested in either writing or contributing to a book, please visit authors.packtpub.com.

Share Your Thoughts

Once you've read *Getting Started with CockroachDB*, we'd love to hear your thoughts! Scan the QR code below to go straight to the Amazon review page for this book and share your feedback.

https://packt.link/r/1-800-56065-6

Your review is important to us and the tech community and will help us make sure we're delivering excellent quality content.

Section 1: Getting to Know CockroachDB

In this section, we will provide a brief introduction to CockroachDB and the motivation behind creating this new database, as well as go into its overall architecture and design concepts.

This section comprises the following chapters:

- *Chapter 1, CockroachDB - A Brief Introduction*
- *Chapter 2, How Does CockroachDB Work Internally?*

1
CockroachDB – A Brief Introduction

In this chapter, we will go over the history of databases, where we will learn about the evolution of SQL, NoSQL, and NewSQL databases, various relational models, different categories for classifying databases, and timelines. Later, we will discuss the CAP theorem. Finally, we will briefly discuss the motivation for creating a new database and learn about the basic architecture of CockroachDB.

The following topics will be covered in this chapter:

- The history and evolution of databases
- Database concepts
- CAP theorem
- CockroachDB

The history and evolution of databases

A database is a collection of data that can be organized, managed, modified, and retrieved using a computer. The system that helps with managing data in a database is called a **database management system** (**DBMS**).

In the 1950s and 1960s, several advancements were made in terms of processors, storage, memory, and networks. We also had our first programming languages, **COBOL** and **FORTRAN**. The development of hard disk drives for data storage further spurred the development of databases. Around the same time, the first notion of a modern-day computer with a mouse and graphical user interface came into existence, making it easy for the general public to consume it. In this section, we will discuss how various types of databases evolved.

SQL

The first database was designed by Charles William Bachman III, an American computer scientist. In 1963, he developed the **Integrated Data Store** (**IDS**), which gave rise to the concept of the navigational database. In **navigational databases**, we can find records by chasing references from other objects. For example, let's say that in a school database, you want to find all the students from a specific grade in a specific school. In a navigational database, first, you have to go to the group of students that belong to a particular school and then to the group that belongs to a particular grade. So, records can be accessed by hierarchical navigation. Based on IDS, Bachman later developed the CODASYL database model in 1969. **CODASYL** stands for **Conference/Committee on Data Systems Languages**, which was a consortium to guide the development of programming languages. Around the same time Edgar F. Codd, an IBM employee, developed the IBM **Information Management System** (**IMS**), which was based on the hierarchical database model. A hierarchical database model is a data model in which the data is designed in a tree-like structure. In 1970, Donald D. Chamberlin and Raymond F. Boyce developed **Structured Query Language** (**SQL**) based on what they'd learned about IMS. They initially called it **Structured English Query Language** (**SEQUEL**), which System R was later developed with by a group at the IBM San Jose research laboratory. In 1976, **QUEL**, which is a relational database query language designed by Michael Ralph Stonebraker, was developed as part of the **Interactive Graphics Retrieval System** (**INGRES**) database management system at the University of California, Berkeley.

Based on QUEL and SQL, several databases were implemented. Some of the most prominent ones include **Post Ingres** (**Postgres**), **Sybase**, Microsoft SQL, **IBM DB2**, Oracle, **MariaDB**, and **MySQL**.

Object-oriented databases

In the 1980s, **object-oriented database systems** (**OODBMSes**) grew in popularity. In OODBMSes, information is represented as objects compared to tables in relational databases. Some of the important ones include `Gemstone/S`, `Objectivity/DB`, `InterSystems Cache`, `Perst`, `ZODB`, `Wakanda`, `ObjectDB`, `ODABA`, and `Realm`.

NoSQL

The concept of non-SQL or non-relational databases has existed since the 1960s, but the term NoSQL became has much more popular in the last decade. NoSQL databases focus on performance and scaling and mostly rely on a non-relational data model such as a document, key-value, wide-column, or graph to organize the data. Some of the most popular ones in this category include **Cassandra, MongoDB, Couchbase, Dynamo, FoundationDB, Neo4j**, and **Hbase**.

NewSQL

With the introduction of the on-demand availability of compute, storage, and network resources and the pay-as-you-go model, which is collectively known as **cloud computing**, the amount of data that we collect, process, manage, and analyze has been growing exponentially. Although it was relatively easier for some of the NoSQL databases to adapt to the cloud, it is still much harder for traditional SQL databases to do so. Many of them are better suited for vertical scaling and do not consider geographically distributed data, the shared-nothing architecture, and enormous scale as part of their initial design. This created a void. We needed SQL databases that are cloud-native, scale well with data growth, and are easy to manage. Many companies developed in-house solutions on top of existing SQL databases:

- **Facebook** developed **TAO**, a NoSQL graph API built on top of sharded MySQL.

- **YouTube** developed **Vitess** to easily scale and manage MySQL clusters.

- **Dropbox** developed **Edgestore**, a metadata store to power their services and products, which again was built on top of MySQL.

- **GreenPlum** developed a massively parallel data platform by the same name for analytics, machine learning, and AI on top of Postgres.

However, it was still relatively hard and painful to manage the data as the underlying database was not built to scale.

In 2012, **Google** published a seminal paper on **Google Spanner**: a globally distributed database service and storage solution. Spanner essentially combined the important features of SQL databases such as ACID transactions, strongly consistent reads, and the SQL interface with some of the features that were only available with NoSQL databases, such as scaling across geographical locations, multi-site replication, and failover. It created a new category of databases called NewSQL, which is meant to indicate a combination of SQL features at NoSQL scale. **YugabyteDB** and **CockroachDB** were developed later, both of which got their inspiration from Google Spanner.

Database concepts

In this section, we will learn about some of the core database concepts, including cardinality, database models, and various processing models.

Cardinality

Before we discuss database models, it is important to know about cardinality. **Cardinality** refers to the relationship between two entities or tables. The most popular ones include *one-to-many*, *many-to-one*, and *many-to-many*.

One-to-one relationship

In the case of a **one-to-one relationship**, a row or entry in one entity or table can be related to only one row in another entity or table. For example, in a Department of Motor Vehicles database, let's say there are two tables called `License Info` and `Driver Info`, as shown in the following diagram:

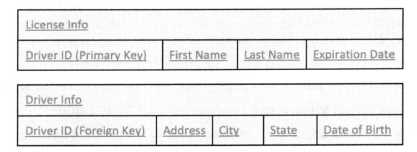

Figure 1.1 – An example of a one-to-one relationship

Here, `Driver ID` can only be assigned to one driver as it has to uniquely identify a driver. Also, a driver can only be assigned one `Driver ID`. So, here, any row in the `License Info` table will be associated with a specific row in the `Driver Info` table.

One-to-many relationship

In a **one-to-many relationship**, a single row from one entity or table can be associated with multiple rows in another entity or table.

For example, let's consider the `Driver Info` and `City Info` tables shown in the following diagram::

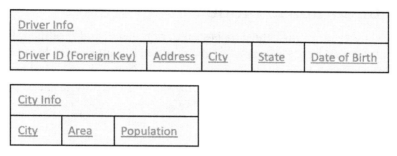

Figure 1.2 – An example of a one-to-many relationship

Here, for every row in `City Info`, there will be multiple rows in `Driver Info`, as there can be many drivers that live in a particular city.

Many-to-many relationship

In a **many-to-many relationship**, a single row in one entity or table can be associated with multiple rows in another entity or table and vice versa.

For example, let's consider two tables: `Vehicle Ownership History`, where we are maintaining the history of ownership of a given vehicle, and `Driver Ownership History`, where we are maintaining the history of vehicles owned by a given driver:

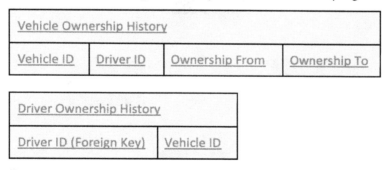

Figure 1.3 – An example of a many-to-many relationship

Here, a driver can own multiple vehicles and a vehicle can have multiple owners over time. So, a given row in the `Vehicle Ownership History` table can be associated with multiple rows in the `Driver Ownership History` table. Similarly, a given row in the `Driver Ownership History` table can be associated with multiple rows in the `Vehicle Ownership History` table.

Now, let's take a look at some of the most important database models.

Overview of database models

A database model determines how the data is stored, organized, and modified. Databases are typically implemented based on a specific data model. It is also possible to borrow concepts from multiple database models when you are designing a new database. The relational database model happens to be the most widely known and has been popularized by databases such as **Oracle**, **IBM DB2**, and **MySQL**.

Hierarchical database model

In the hierarchical database model, the data is organized in the form of a tree. There is a root at the first level and multiple children at the subsequent levels. Since a single parent can have multiple children, one-to-many relationships can easily be represented here. A child cannot have multiple parents, so this results in the advantage of not being able to model many-to-many relationships.

IBM's **Information Management System (IMS)** was the first database that implemented this data model.

The following diagram shows an example of a hierarchical database model:

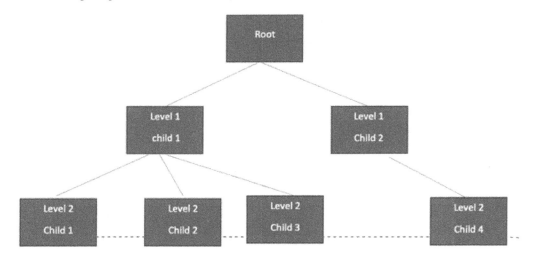

Figure 1.4 – An example of a hierarchical database model

Typically, the tree starts with a single root and the data is organized into this tree. Any node except the leaves can have multiple children, but a child can have only one parent.

Network model

The network model was developed as an enhancement of the hierarchical database model to accommodate many-to-many relationships. The network model relies on a graph structure to organize its data. So, there is no concept of a single root, and a child can have multiple parents and a parent can have multiple children. **Integrated Data Store (IDS)**, **Integrated Database Management Systems (IDMS)**, and **Raima Database Manager (RDM)** are some of the popular databases that use the network model.

As shown in the following diagram, there is no single root and a given child (for example, *Object 2* can have multiple parents; that is, *Object 1* and *Object 3*):

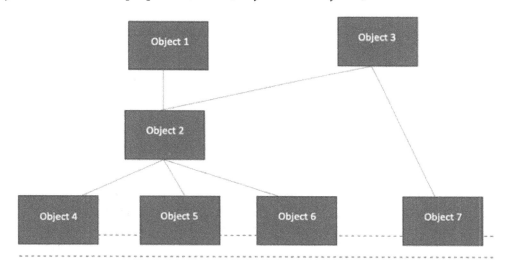

Figure 1.5 – An example of a network model

Relational model

Although the network model was an improvement over the hierarchical model, it was still a little restrictive when it came to representing data. In the relational model, any record can have a relationship with any other with the help of a common field. This drastically reduced the design's complexity and made it easier to independently add, update, and access records, without having to walk down the tree or traverse the graph. SQL was combined with the relational database model to provide a simple query interface to add and retrieve data.

All the popular traditional databases such as Oracle database, IBM DB2, MySQL, MariaDB, and Microsoft SQL Server implement relational data models.

Let's look at two tables called `Employee` and `Employee Info`:

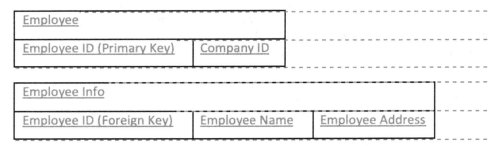

Figure 1.6 – Employee tables showing the column names

Here, `Employee ID` is the common field or column between the `Employee` and `Employee Info` tables. The `Employee` table is responsible for ensuring that a given `Employee ID` is unique, while `Employee Info` is responsible for more detailed information about a given employee.

Object-relational model

The **object-relational model**, as the name suggests, combines the best of the relational and object data models. The concept of objects, classes, and inheritance are directly supported as first-class citizens as part of the database and in queries. SQL:1999, the fourth revision of SQL, introduced several features for embedding object concepts into the relational database. One of the main features was to create structured user-defined types with `CREATE TYPE` to define an object's structure.

Over time, relational databases have added more support for objects. There is a varying degree of support for object concepts in Oracle database, IBM DB2, PostgreSQL, and Microsoft SQL Server.

Given the scope of this book, we will not discuss the **entity-relational model**, **object model**, **document model**, **star schema**, **snowflake schema**, and many other less well-known models.

Now, let's look at how databases can be classified based on what kinds of workload they can be used for.

Processing models

Based on how you want to consume and process data, databases can be categorized into four different processing systems. Let's take a look.

Online transaction processing (OLTP)

OLTP systems support the concept of transactions. A transaction refers to the ability to atomically apply changes (insert, update, delete, and read) to a given system. One popular example is a bank, where withdrawing or depositing money to a given bank account must be done atomically to ensure data is not lost or incorrect. So, the main purpose here is to maintain data integrity and consistency. Also, these systems are generally suited for fast-running queries.

Online analytical processing (OLAP)

OLAP focuses mostly on running queries to analyze multi-dimensional data and to extract some intelligence or patterns from it. Typically, such systems support generating some sort of report that can be used for marketing, sales, financing, budgeting, management, and many more. Data mining and data analytics applications would typically have to have an OLAP system in some form. OLAP doesn't deal with transactions, and the emphasis is more on analyzing large amounts of data from different sources to extract business intelligence. Some databases also provide built-in support for MapReduce to run queries across a large set of data.

A data warehouse is a piece of software that's used for reporting and data analysis. Warehouses are typically developed for OLAP. It is also very common to retrieve the data from OLTP in batches or bulk, run it through an **Extract, Load, and Transform** (**ELT**) or **Extract, Transform, and Load** (**ETL**) data transformation pipeline, and store it in an OLAP system.

Online event processing (OLEP)

OLEP guarantees strong consistency without the traditional atomic commit protocols or distributed locking. OLEP also focuses on high performance, larger scales, and fault tolerance.

Hybrid transaction/analytical processing (HTAP)

As the name suggests, this system tries to provide the best of both transactions and analytical processing. Most of the NoSQL and NewSQL databases provide support for managing both transactional and analytical workloads. Vitess is a database clustering system that can be used to scale and shard MySQL instances. Vitess provides HTAP features on top of MySQL by allowing a given MySQL instance to be configured as master or read-only, where read-only can be used for analytical queries and MapReduce. It is possible to use CockroachDB as **HTAP** by propagating changes with the help of **change data capture** (CDC) in the OLTP cluster or primary cluster to a separate cluster, which is solely used for analytical processing.

Now, let's learn a bit about embedded and mobile databases, including why they exist and some of the most popular ones in this space.

Embedded and mobile databases

Embedded databases usually refer to databases that can be tightly integrated into an application, without needing separate hardware to support them. Also, they don't have to be managed separately. Some of the most popular embedded databases include **SQLite**, **Berkeley DB** from Oracle Corporation, and **SQL Server Compact** from Microsoft Corporation. Embedded databases are also very useful for testing purposes as they can be started within test suites.

Mobile database refers to the class of databases that work with very limited memory footprint and compute and can be deployed within a mobile device. They are typically used for storing user data for apps running on mobile devices. SQLite, SQL Server Compact, **Oracle database Lite**, **Couchbase Lite**, **SQL Anywhere**, **SQL Server Express**, and **DB2 Everyplace** belong to this category,

Database storage engines

A database storage engine is a component within a database management system that is responsible for **Create, Read, Update, Delete** (CRUD) operations and transferring data between disk and memory, without compromising data integrity. Some of the most popular ones include **Apache Derby**, **HSQLDB**, **InfinityDB**, **LevelDB**, **RocksDB**, and SQLite. CockroachDB initially started with RocksDB as its database engine, but from release 20.2 onward, **Pebble** will be the database engine by default. Pebble, as per Cockroach Labs, is a RocksDB-inspired and RocksDB-compatible key-value store focused on the needs of CockroachDB. RocksDB was implemented in **C++**, whereas Pebble was implemented in **Golang**. This makes it easier to manage and maintain as CockroachDB itself was written in Golang. This means that we only have to deal with one language now.

CAP theorem

Eric A. Brewer gave a keynote talk in 2000 titled *Towards Robust Distributed Systems* at a symposium on *Principles of Distributed Computing*, summarizing his years of learning about distributed systems. Brewer talked about key aspects of a distributed system: consistency, availability, and tolerance toward network partition. Consistency refers to the fact that every read should see the data from the most recent write; otherwise, it should error out. Availability means every requested read or write should receive a non-error response. Partition tolerance indicates that the system should continue to serve, irrespective of delays and communication failures between nodes in the system. **Consistency, Availability, and Partition Tolerance (CAP)** theorem claims that, at most, you can only have two of these three properties in a distributed system.

Consistency and partition tolerance (CP)

A **CP database** provides consistency and partition tolerance but cannot provide availability. This is also called a **CAP-consistent system**. Let's understand this by looking at an example:

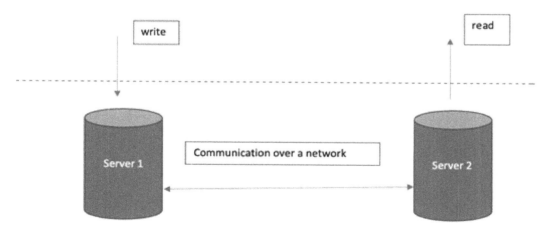

Figure 1.7 – CP system

Let's consider the system shown in the preceding diagram, where two servers are serving read and write traffic. For this example, let's say writes only land on **Server 1** and reads only land on **Server 2**. So long as **Server 1** can talk to **Server 2**, all the writes that come to **Server 1** can be propagated synchronously to **Server 2**. This ensures that any reads that come to **Server 2** are always consistent, which means they see the latest data written by the latest write in **Server 1**:

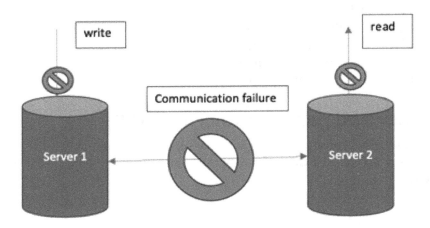

Figure 1.8 – CP system during a communication failure

Now, let's say that, as shown in the preceding diagram, the communication between **Server 1** and **Server 2** has broken down and now **Server 1** is no longer able to propagate the writes synchronously. This results in partitioning. Since the data cannot be propagated between the two servers, read or write traffic cannot be served until we resolve the partition issue as we have to ensure data consistency.

Some of the most popular databases that have CP characteristics are HBase, Couchbase, and MongoDB. CockroachDB also falls into this category.

Availability and partition tolerance (AP)

In this case, a database is guaranteed to always be available and it can tolerate partitioning, but at the cost of consistency. This is also known as a **CAP-available system**. Here, the application is expected to deal with data consistency:

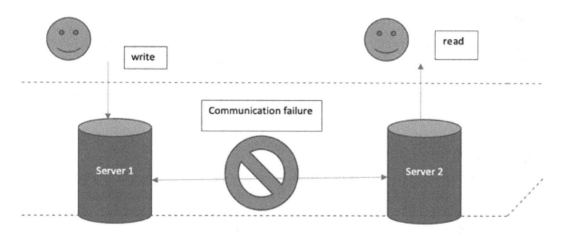

Figure 1.9 – AP system during a communication failure

Similar to the previous example, if the communication between **Server 1** and **Server 2** breaks down, **Server 1** and **Server 2** continue to serve the traffic but reads to **Server 1** and **Server 2** might return different versions of the data, based on when the communication has failed and whether there was any change to that data, after the communication failure. **Cassandra**, **Riak**, and **CouchDB** are popular examples of AP databases.

Consistency and availability (CA)

In the case of a **CA database**, the system cannot tolerate partitioning but can guarantee consistency and availability. Traditional databases with single-server deployments with no replication or slaves can be classified as CA. Now, many traditional RDBMS databases can be configured in various ways to have CA, CP, or AP as desired.

CockroachDB

The name CockroachDB was inspired by the insect that goes by the same name. Just like how cockroaches have been surviving for millions of years and colonizing the entire planet and thriving, CockroachDB instances are supposed to replicate and repair data, spread naturally across multiple availability zones, and survive total regional failures. Also, once CockroachDB becomes part of a given software ecosystem, it's impossible to get rid of or replace it, just like cockroaches. Here, we will discuss why there is a need for yet another database, known as Inspiration, and provide a high-level overview of CockroachDB.

Why yet another database?

As more companies shift from on-premises to the cloud, they are looking for SQL datastores on various cloud platforms to manage their transactional data. Most of the traditional databases such as MySQL, Postgres, and Oracle are not built for the cloud. This necessitates a cloud-native, consistent, distributed SQL that can scale with the growth of data. CockroachDB fills this gap.

Inspiration

As we previously discussed in the *NewSQL* section, in 2012, Google published a seminal paper on Google Spanner: a globally distributed database service and storage solution. Although Google Spanner combined the best of both SQL and NoSQL and was very useful for a lot of applications, it was not available for public usage. Also, Google Spanner was and still is not an open source project and has only been available on Google Cloud Platform since 2017. So, this created a necessity for an open source Spanner-like database that can be used in different cloud providers and on-premises. Around 2012, Spencer Kimball, Peter Mattis, and Ben Darnell were working at Google on the Google File System and Google Reader projects. They also got acquainted with both **Bigtable** and Spanner during their tenure at Google. They decided to build something very similar to Spanner to make it available for everyone and started an open source project on **GitHub** in 2014. After a year, they decided to leave Google and founded Cockroach Labs in 2015 before officially working on CockroachDB in June 2015.

Key terms and concepts

Before we look at the various functional layers, let's look at some of the key concepts and terms. A CockroachDB **cluster** refers to a group of nodes that act as a single logical unit. A **node** is a single machine that runs an instance of CockroachDB. CockroachDB stores all the data as sorted key-value pairs. These keys are divided into **ranges**. CockroachDB replicates each range and stores each replica on a different node. For each range, there will be a **leaseholder**, which acts as a primary owner of a given range and receives and coordinates all the traffic for that range. For each range, one of the replicas acts as a **leader** for write requests and ensures that the majority of the replicas are in consensus, before committing a given write. For each range, there will be a time-ordered log of writes, called a **raft log**, for which the majority of replicas agreed upon.

High-level overview

CockroachDB is a cloud-native, consistent, highly scalable relational database. Some of the primary goals of CockroachDB are to provide strong consistency, geo-distribution of data, high availability, SQL support, easy deployment, and less maintenance. Since we will be dealing with CockroachDB internals in detail in subsequent chapters, we will just provide a high-level overview here:

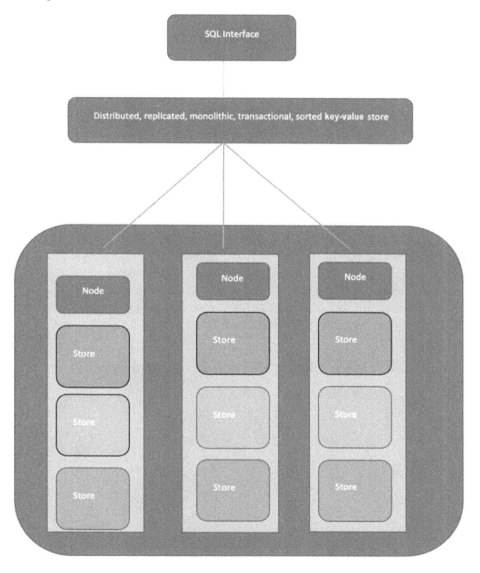

Figure 1.10 – High-level overview of the CockroachDB architecture

CockroachDB exposes a SQL interface, using which clients can interact with the database. Client requests can land on any node within a given cluster and work just fine since all the nodes are symmetrical.

CockroachDB can be divided into five functional layers:

- **SQL**
- **Transactional**
- **Distribution**
- **Replication**
- **Storage**

The **SQL layer** is responsible for receiving SQL queries and converting them into key-value operations. The **transactional layer** ensures that all CRUD operations that happen on multiple key-value pairs are transactional. The **distribution layer** is responsible for ensuring ranges are evenly distributed among all the available nodes in a cluster. The **replication layer** ensures that ranges are replicated synchronously, whenever there is a change. Finally, the **storage layer** is responsible for managing key-value data on the disk.

Summary

In this chapter, we learned about the evolution of databases, how databases can be categorized based on various criteria, CAP theorem, and a brief introduction to CockroachDB. By now, you should also be familiar with database and processing models, what the CP, CA, and AP systems in CAP theorem offer, and the functional layers of CockroachDB.

In the next chapter, we will take a deep dive into CockroachDB's architecture and design concepts.

2
How Does CockroachDB Work Internally?

In the previous chapter, we learned about the evolution of databases and the high-level architecture of **CockroachDB**. In this chapter, we will go a bit deeper into each of the layers of CockroachDB and explore how CockroachDB works internally. We will also discuss some of the core design aspects that form the basic pillars of CockroachDB.

CockroachDB can be broadly divided into five main layers, as outlined here:

- Structured Query Language (SQL)
- Transactional
- Distribution
- Replication
- Storage

Each of these layers will be explained as the main topics of this chapter, in the following order:

- Installing a single-node CockroachDB cluster using Docker
- Execution of a SQL query
- Managing a transactional key-value store
- Data distribution across multiple nodes
- Data replication for resilience and availability
- Interactions with the disk for data storage

Since we will be trying out some commands in this chapter, it's important to have a working environment for them. So, we will start with the technical requirements, where we will go over how to set up a single-node CockroachDB cluster.

Technical requirements

To try out some of the commands in this chapter, you will need a single-node CockroachDB cluster. There are several ways of installing CockroachDB on a computer, as outlined here:

- Use a package manager such as Homebrew to install it, but this option only works on a Mac.
- Download binaries, extract it, and set it in the PATH variable.
- Use **Kubernetes** to orchestrate CockroachDB pods.
- Build from source and install.
- Download a **Docker** image and run it.

In the current chapter, we will just go over how to run CockroachDB using Docker since the steps are common, irrespective of the operating system that you are using.

To use CockroachDB with Docker, you need a computer with the following:

- At least 4 **gigabytes** (**GB**) of **random-access memory** (**RAM**)
- 250 GB of disk space
- Docker installed

In the next section, we will learn about installing CockroachDB.

Installing a single-node CockroachDB cluster using Docker

Let's take a look at how to install a single-node CockroachDB cluster, which will be required to try out some of the commands that will be introduced in this chapter. Here are the steps you need to follow:

1. Ensure the Docker daemon is running with the following command:

   ```
   docker version
   ```

   Pull the most recent stable version with tag `v<xx.y.z>` from `https://hub.docker.com/r/cockroachdb/cockroach/`. Take a look at the following example:

   ```
   docker pull cockroachdb/cockroach:v20.2.4
   ```

2. Make sure this image is available and the version is correct with the help of the following command:

   ```
   docker images | grep cockroach
   cockroachdb/cockroach    v20.2.4              d47481b0b677
   2 days ago          329MB
   ```

   If running docker on windows, then replace grep with findstr:

   ```
   docker images | findstr cockroach
   ```

3. Create a **bridge network**. A bridge network allows multiple containers to communicate with each other. Here's the code you'll need to create one:

   ```
   $ docker network create -d bridge crdb_net
   ```

4. Create a volume. Volumes are used for persisting data generated and used by Docker containers. Since CockroachDB is a database, it needs a place to store the data, and hence we should attach a volume to the container. You can do this by running the following command:

   ```
   $ docker volume create crdb_vol1
   ```

5. Start a CockroachDB node using the following command:

```
$ docker run -d \
--name=crdb1 \
--hostname=crdb1 \
--net=crdb_net \
-p 26257:26257 -p 8080:8080  \
-v "crdb_vol1:/cockroach/cockroach-data"  \
cockroachdb/cockroach:v20.2.4 start \
--insecure \
--join=crdb_vol1
```

- Here is an explanation of what each of these options means:
- docker run: This starts a Docker container.
- --name: Name of the container.
- --hostname: Hostname of the container. This will be useful if you are running multiple containers and want to join them to form a cluster.
- --net: Bridge network. This will be useful if you have more than one container that wants to communicate with other containers.
- -p 26257:26257: Port mapping for inter-node or SQL client for node communication.
- -p 8080:8080: Port mapping used for **HyperText Transfer Protocol** (HTTP) requests to the CockroachDB console.
- -v "crdb_vol1:/cockroach/cockroach-data": Mounts the host directory as a data volume.
- cockroachdb/cockroach:v20.2.4 start: Command to start the CockroachDB node.
- --insecure: Option to start the node in insecure mode.
- --join=crdb_vol1: Here, you can specify multiple hostnames of CockroachDB nodes that will form a cluster. For the current chapter, we just need a single-node cluster.

6. Initialize the cluster with the following command:

```
$ docker exec -it crdb1 ./cockroach init --insecure
Cluster successfully initialized
```

7. To ensure that the CockroachDB node is functional, we can create a database and a table, insert some data, and run a query.

    The following command is used for starting the SQL shell:

    ```
    docker exec -it crdb1 ./cockroach sql --insecure
    ```

    The following command is used for creating a database:

    ```
    root@:26257/defaultdb> CREATE DATABASE testdb;
    ```

    The following command is used for creating a table:

    ```
    root@:26257/defaultdb> CREATE TABLE testdb.testtable (id
    INT PRIMARY KEY, string name);
    ```

8. Verify the columns by running the following code:

    ```
    root@:26257/defaultdb> SHOW COLUMNS from testdb.
    testtable;
      column_name | data_type | is_nullable | column_default
    | generation_expression |  indices  | is_hidden
    ---------------+-----------+-------------+---------------
    +----------------------+-----------+------------
        id         | INT8      |     false   | NULL
    |                      | {primary} |    false
        string     | NAME      |     true    | NULL
    |                      | {}        |    false
    (2 rows)


    Time: 134ms total (execution 133ms / network 1ms)
    ```

9. Insert some data, as follows:

    ```
    root@:26257/defaultdb> INSERT INTO testdb.testtable
    VALUES (1,'Spencer Kimball'), (2,'Ben Darnell'),
    (3,'Peter Mattis');
    ```

    Run a query to fetch the contents of the table:

    ```
    root@:26257/defaultdb> SELECT * FROM testdb.testtable;
      id |       string
    -----+-------------------
       1 | Spencer Kimball
       2 | Ben Darnell
       3 | Peter Mattis
    ```

```
(3 rows)
```

```
Time: 5ms total (execution 3ms / network 2ms)
```

Next, we will learn about each of the layers, starting with the SQL layer.

Execution of a SQL query

Any application that talks to CockroachDB can use Postgres-compatible drivers to talk to CockroachDB. If you prefer **object-relational mappers (ORMs)**, **Golang ORM (GORM)**, go-pg, and **SQLBoiler** can be used to interact with CockroachDB. Irrespective of where the actual data resides, you can issue a query to any of the nodes in the cluster. Whichever node the query lands in, that node acts as a gateway.

Requests from SQL clients are received as **SQL statements**. The SQL layer is responsible for converting these SQL statements into a plan of key-value operations and passing it to the transaction layer. CockroachDB, as of version 21.2, supports both native drivers and the **Postgres** wire protocol. A wire protocol defines how two applications can communicate over a network.

CockroachDB has support for most **American National Standards Institute** (**ANSI**) SQL standards to change table structures and data. Next, we will look at the various stages of query execution.

SQL query execution

As with any other database, there are standard steps involved in processing incoming SQL requests and serving the data.

Before going through the different steps of a SQL layer, it's important to get ourselves familiar with some terms related to SQL, as follows:

- An **SQL parser** is software that scans a given SQL statement in its string form and tries to make sense of it. This involves **lexical analysis**, which is extracting the tokens or the keywords, and syntactic analysis, where you make sure the entire query is valid and can be represented in a form that makes it easier to execute the query. Lex is a popular lexical analyzer and **Yacc** (which stands for **Yet Another Compiler-Compiler**) is software that processes grammar and generates a parser.

- An **abstract syntax tree** (**AST**) or a syntax tree is a tree representation of a given domain-specific language. Nodes in the tree represent constructs in that language. ASTs are essential for deciding on how to execute a given query.

Some of the main steps in a SQL layer include query parsing, logical planning, physical planning, and query execution. Vectorized query execution is enabled by default, which gives much better performance.

Parsing

CockroachDB uses **Goyacc**, Golang's equivalent of the famous **C Yacc** to generate a SQL parser from a grammar file located at `pkg/sql/parser/sql.y`. This parser then converts the input into an AST comprising of tree nodes, where the node types are defined under the `pkg/sql/sem/tree` package.

Logical planning

Once we have an AST, it then has to be transformed into a **logical plan**. A logical plan defines how various clauses can be logically ordered. As part of this transformation, semantic analysis is initially done for type checking, name resolution, and to ensure the query is valid. Later, this logic plan is simplified without changing the overall semantics and is optimized based on the cost. Here, cost refers to the total time taken to return the results of a query. Cost optimization involves picking the right indexes, query optimization, and selecting the best strategies for sorting and joining. We can view the logical plan using `EXPLAIN <SQL Statement>`.

The following example shows the output of a sample `EXPLAIN` query:

```
root@:26257/defaultdb> EXPLAIN SELECT * FROM testdb.testtable;
   tree  |          field         |       description
---------+------------------------+---------------------
         | distribution           | full
         | vectorized             | false
   scan  |                        |
         | estimated row count    | 1
         | table                  | testtable@primary
         | spans                  | FULL SCAN
(6 rows)


Time: 11ms total (execution 10ms / network 1ms)
```

Physical planning

In this phase, all the participating nodes are determined based on where the data resides and also who is the primary owner for a given range, which is also known as a range's **leaseholder**. Typically, partial result sets are gathered from various nodes that are then sent to the coordinator or the gateway node, which aggregates all the results and sends a single response back to the SQL client. Let's look at an example of what this looks like for a sample query for the `testdb.testtable` table that we created in *step 8 in the Installing a single node CockroachDB cluster using Docker* section.

Let's assume we have six names that are distributed across four nodes in a cluster. The following table shows information about the leaseholder nodes, key ranges, and user data:

Leaseholder node	Key range	Names
node 1	[A – G)	Ben
		Darnell
node 2	[G – L)	Jordan
		Kimball
node 3	[L – Q)	Peter
		Mattis
node 4	[Q – Z]	Spencer
		Zoey

Figure 2.1 – Table showing information about key ranges

Now, let's assume that a SQL client sends a query to fetch all the names from the `testdb.testtable` table. In this example, this request lands in **node 1**, so it acts as a gateway node that is responsible for collecting the data required to serve this request after coordinating with all relevant leaseholder nodes, getting a partial result, aggregating the data, and sending it back to the SQL client. This process is illustrated in the following screenshot:

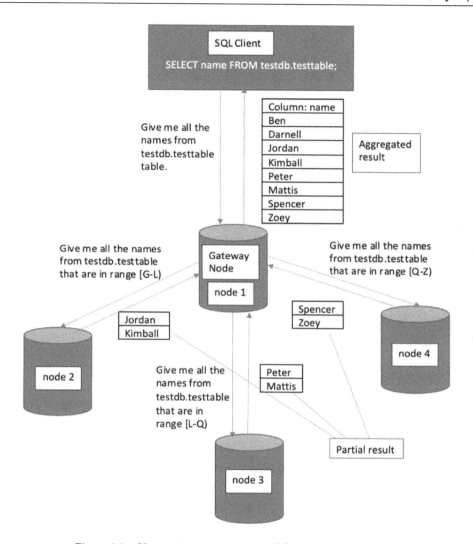

Figure 2.2 – How a given query is served from the gateway node

At **Gateway Node**, the main challenge is to identify whether a given computation should be pushed down to nodes where the data resides or to do the same computation within the coordinator node on an aggregated result. The physical plan aims to make the best use of parallel computing and at the same time reduce the overall data that gets transferred between data and coordinator nodes. Also, table joins bring in an added layer of complexity. We can view the physical plan using EXPLAIN(DISTSQL) <SQL statement>. DISTSQL generates a **Uniform Resource Locator** (**URL**) for a physical query plan, which includes high-level information about how the query will be executed.

The following example shows the output of a sample EXPLAIN(DISTSQL) query:

```
root@:26257/defaultdb> EXPLAIN(DISTSQL) SELECT * FROM testdb.
testtable where id > 1;
  automatic |                                                           url
------------+-------------------------------------------------------------------
------------------------------------------------------------------------------
------------------------------------------------------------------------------
------------------------------------------------------------------------------
------------------------------------------------------------------------------
------------------------------------------------------------------------------
------------------------------------
    false   | https://cockroachdb.github.io/distsqlplan/decode.
html#eJyMj0FLxDAQhe_-ivBOKtFt9ZaTohULdXdtCwraQ7YZlkK3qZkUl
NL_Lm1A8SDsKbz3Mu-bGcEfLRSS1212m67F6X1alMVzdiaKJEvuSnEuHvLNk_
DE3uwu58frXUvi5THJE9EY8T5E0TWJGBKdNbTWB2KoN8SoJHpna2K2brbG5UNq
PqEiiabrBz_blURtHUGN8I1vCQr1jMhJG3KrCBKGvG7apfZng5
veNQftviBR9LpjJVZXF6gmCTv432r2ek9Q8SSPx-fEve2Y_pD_
a46mSoLMnsKJbAdX09bZesEEuVnmFsMQ-5DGQaRdiKZqOvkOAAD__6E0gSw=
(1 row)
```

```
Time: 21ms total (execution 16ms / network 4ms)
```

As you can see, the output is a downloadable URL for the actual physical plan. If you click on that URL, it will take you to the physical plan, as shown in the following screenshot:

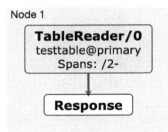

Figure 2.3 – Visual representation of a physical plan in a single-node cluster

The EXPLAIN ANALYZE statement (https://www.cockroachlabs.com/docs/
v21.1/sql-statements) executes a SQL query and generates a statement plan with execution statistics, as illustrated in the following code snippet:

```
root@:26257/testdb> EXPLAIN ANALYZE SELECT * FROM testdb.
testtable;
  automatic |                                                           url
```

```
-----------------+-----------------------------------------------
-----------------------------------------------------------------
-----------------------------------------------------------------
-----------------------------------------------------------------
-----------------------------------------------------------------
-----------------------------------------------------------------
-----------------------------------------------------------------
-----------
     true    | https://cockroachdb.github.io/distsqlplan/decode
.html#eJyMUE1L60AU3b9fMZzVe4-xNhZdzMqqEQKxrU0XfpDFNHMpgSQT596i
peS_SxJUXAiuhvMx5xzuEfxSwSB-WKXzZKHmi3n6-BSrvzdJtsnu038qi9P4eq
P-q9v18k4JsbjtpH_EbiuCRuMdLWxNDPOMCLlGG3xBzD701HEwJO4NZqpRNu1ee
jrXKHwgmCOk1IpgsOkD12QdhdMpNByJLash9rPvsg1lbcMBGl1rGzbq
BBrBv7IKZJ1RM2iw2KpSUtZkVDSZXcxqhsb2IPThis7O1RXyTsPv5WsRi90RTNT
p369eE7e-Yfo2-KfkaZdrkNvReBn2-1DQKvhiqBnhcvg3EI5YRjUaQdKMUpd3f9
4DAAD__yWSjvw=
(1 row)
```

```
Time: 4ms total (execution 3ms / network 1ms)
```

The output is shown here:

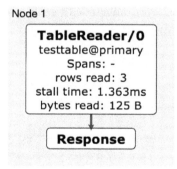

Figure 2.4 – Statement plan with execution statistics

EXPLAIN ANALYZE (DEBUG) executes a query and generates a link to a ZIP file that contains the physical statement plan (https://www.cockroachlabs.com/docs/stable/explain-analyze.html#distsql-plan-viewer), execution statistics, statement tracing, and other information about the query. The code is illustrated in the following snippet:

```
root@:26257/testdb> EXPLAIN ANALYZE (DEBUG) SELECT * FROM
testdb.testtable;
                            text
```

```
-----------------------------------------------------------------
-----------------
  Statement diagnostics bundle generated. Download from the
Admin UI (Advanced
  Debug -> Statement Diagnostics History), via the direct link
below, or using
  the command line.
  Admin UI: http://crdb1:8080
  Direct link: http://crdb1:8080/_admin/v1/
stmtbundle/676836398188756993
  Command line: cockroach statement-diag list / download
(6 rows)


Time: 92ms total (execution 91ms / network 1ms)
```

Query execution

During query execution, the physical plan is pushed down to all the data nodes that would be involved in serving a given query. CockroachDB uses work units called **logical processors**, which will be responsible for executing relevant computations. Logical processors across data nodes also communicate with each other so that data can be sent back to the coordinator or the gateway node.

There are two types of query execution, as outlined here:

- Non-vectorized or row-oriented query execution
- Vectorized or column-oriented query execution

Vectorized query execution (**column-oriented query execution**) is more suited to analytical workloads, whereas **non-vectorized query execution** (**row-oriented query execution**) is preferred for transactional workloads. By default, vectorized query execution is enabled on CockroachDB.

Now, let's take a look at how CockroachDB provides a transactional key-value store.

Managing a transactional key-value store

A **transactional layer** involves implementing a concurrency control protocol, as multiple transactions can try to update the same data at the same time, which can result in a conflict. In concurrency control, there are two different ways of dealing with conflicts, as outlined here:

- Avoid conflicts altogether with pessimistic locking—for example, a read/write lock.

- Let the conflict happen but detect it with optimistic locking and resolve it—for example, **multi-version concurrency control (MVCC)**.

CockroachDB uses MVCC. In MVCC, there can be multiple versions of the same record, but you resolve the conflicts before committing the changes.

In CockroachDB, a given transaction is executed in three phases, outlined next:

1. A transaction is started with a target range that will participate in the transaction. A new transaction record is created to track the status of the transaction. It will have the initial state as PENDING. At the same time, a write intent is created. In CockroachDB, instead of directly writing the data to the storage layer, data is written to a provisional state called write intent. Here, the intent flag indicates that the value will be committed once the transaction is committed. CockroachDB uses MVCC for concurrency control. The transaction **identifier (ID)** is used to resolve conflicts for write intents. Each node involved in the transaction returns the timestamp used for the write, and the coordinator node selects the highest timestamp among all write timestamps and uses it in the final commit timestamp.

2. The transaction is marked as committed by updating its transaction record. The commit value also contains the candidate timestamp. The candidate timestamp is a temporary timestamp to denote when the transaction is committed and is selected as the actual node coordinating the transaction. Once the transaction is completed, control is returned to the client.

3. After the transaction is committed, all written intents are updated in parallel by removing the intent flag. The transaction coordinator does not wait for this step to be completed before returning the control to the client.

We will learn more about conflict resolution, **atomicity, consistency, isolation, and durability (ACID)**, logical clocks, and transaction management in the next chapter. Next, we will go over the distributed layer.

Data distribution across multiple nodes

A table in CockroachDB can be partitioned, and this is discussed in *Chapter 4, Geo-Partitioning*, where we talk about geo-partitioning. CockroachDB stores the data in a **monolithic sorted key-value store (MSKVS)**. **Key-space** is all the data you have in a given cluster, including information about its location. Key-space is divided into contiguous batches, called **ranges**. The MSKVS makes it easy to access any data from any node, which makes it possible for any node in the cluster to act as a gateway node, coordinating one or more data nodes while serving client requests.

The MSKVS

The MSKVS contains two categories of data, as outlined here:

- **System data**, which contains meta ranges, where the data of each range can be found within the cluster.

- **User data**, which is the actual table data.

Meta ranges

The location of ranges is maintained in two-level indexes, known as **meta ranges**. The first level (a.k.a. meta1) points to the second level (a.k.a. meta2) and the second-level indexes point to the actual data. This is shown in the following screenshot:

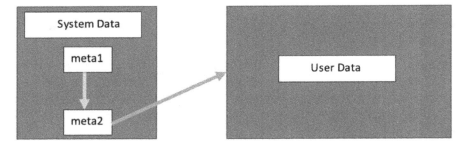

Figure 2.5 – Meta-range management in two levels

Every node in the cluster has complete information of meta1 and that range is never split. meta2 data is cached on nodes. These are invalidated whenever ranges change, and the cache gets updated with the latest value.

Let's look at an example.

Here, we will understand what meta1 and meta2 data looks like for an alphabetically sorted column. When we write ranges, square brackets '[' and ']' indicate that the number is included in the range, and parentheses '(' and ')' mean that the number is excluded.

Let's look at some examples here:

- [1,10] —range starts from 1 and ends at 10 as both numbers are included
- [1,5) —range starts from 1 and ends at 4, as 5 is excluded
- (1, 8) —range starts from 2, as 1 is excluded and it ends at 7, as 8 is excluded

Let's now understand what meta1 and meta2 look like for an alphabetically sorted column, as follows:

1. meta1 contains addresses of nodes that contain meta2 replicas. Let's assume there are two meta1 entries for simplicity.

   The first meta1 entry points to the meta2 range for keys [A-M), and the second meta1 entry points to the meta2 range for keys [M-Z]. Here, maxKey indicates the rest of the range till the maximum available key. Since we are talking about an alphabetically sorted column, that would start at M, as the previous range excluded M and ends at Z, which is the last letter of the alphabet. The code is illustrated here:

   ```
   meta1/M -> node1:26257, node2:26257, node3:26257


   meta1/maxKey -> node4:26257, node5:26257, node6:26257
   ```

2. meta2 contains addresses of the nodes containing actual user data for that alphabetically sorted column. The first entry in the value always refers to the leaseholder, which is the primary owner of a given range. The code is illustrated here:

   ```
   meta2 entry for the range [A-G)
   meta2/G -> node1:26257, node2:26257, node3:26257

   meta2 entry for the range [G-M)
   meta2/M -> node1:26257, node2:26257, node3:26257

   meta2 entry for the range [M-Z)
   meta2/Z -> node4:26257, node5:26257, node6:26257

   meta2 entry for the range [Z-maxKey)
   meta2/maxKey->  node4:26257, node5:26257, node6:26257
   ```

Table data

When a new table is created, the table and its secondary indexes all point to a single range. Once the range size exceeds 512 **megabytes (MB)**, the range is split into two. This continues as the data grows. Table ranges are replicated to multiple nodes for survivability so that even if some nodes in the cluster shut down or crash, there will not be any data loss.

Next, we will learn about replication and Raft, a distributed consensus algorithm.

Data replication for resilience and availability

This layer is responsible for ensuring that the table data is replicated to more than one node and also keeps the data consistent between replicas.

The **replication factor** indicates how many replicas of a specific table's data should be kept—for example, if the replication factor is 3, CockroachDB keeps three copies of all the table data. The number of node failures that can be tolerated without data loss = (replication factor – 1) / 2; for example, if the replication factor is 3, then (3 – 1) / 2 = 1 node failure can be tolerated. Whenever a node goes down, CockroachDB automatically detects it and works toward making sure the data in the node that went down is replicated to other nodes, in order to honor the replication factor and also to increase survivability.

CockroachDB uses the Raft distributed consensus algorithm, which ensures a quorum of replicas agree on changes to ranges before those changes are committed.

What is consensus?

Consensus is a concept in distributed systems that is used for fault-tolerance and reliability when some nodes either go down or will not be reachable because of network issues. Consensus involves multiple nodes agreeing to changes before they are committed. If all the nodes involved in the consensus are not available, then an agreement can still be made, as long as a majority of the nodes are available—for example, in a cluster of nine nodes, we need at least five nodes that are able to communicate with each other in order to reach an agreement. Paxos, Multi-Paxos, Raft, and Blockchain are some of the popular consensus algorithms.

The Raft distributed consensus protocol

The word **Raft** is supposed to be the combination of *R* (which stands for *reliable*, *replicated*, and *redundant*), A (which stands for *and*), and *FT* (which stands for *fault tolerance*). Although it's not an acronym, the word *Raft* is supposed to be a system that provides reliability, replication, redundancy, and fault tolerance.

In Raft, all nodes that have a replica of a given range will be part of a Raft group. Each node can be in one of the following states:

- **Leader**—Acts as a leader of the Raft group. Responsible for managing data mutations and ensuring that data is consistent between the leader and its followers using log replication.

- **Follower**—Follows a leader and works with the leader in order to keep the data consistent.

- **Candidate**—In the absence of a clear leader, any participating node can try to become a leader. A node that is trying to become a leader is called a candidate.

There are mainly two types of **remote procedure call** (**RPC**) requests, as listed here:

- `RequestVote`—Used for requesting votes by candidate node to other participating nodes

- `AppendEntries`—Used for log replication and heartbeat

Let's now understand how leader election happens within Raft.

Leader election

Initially, all nodes of a Raft group start as followers. If they don't hear from a leader, they become a candidate. The candidate votes for itself and requests votes by sending a `RequestVote` message to other participating nodes. Any candidate with a majority of votes becomes the leader. This process is called *leader election*. If two nodes end up with the same number of votes, then there will be *re-election*. Also, election timeout is randomized among the nodes of a Raft group, which ensures each participating node becomes a candidate at different points of time. This reduces the chance of a split vote.

After the election, the leader keeps sending heartbeats through an `AppendEntries` message to all its followers, and the followers keep responding. This ensures that the election term is maintained.

There will be re-election when one or more nodes don't hear from the leader for a certain period of time. This is called *election timeout*.

Log replication

Once a leader is elected, all changes to the data go through the leader. Every change is first recorded in the leader node's log. The actual node's value is not updated till the change is committed. In order to commit the change, the leader first broadcasts it via an AppendEntries message to all its followers. After this, the leader waits for a majority of the nodes to reply back. The followers respond back once they make an entry in their own logs. Once the leader receives a response from a majority of the nodes, it then commits the change and modifies the node value. After this, the leader notifies all the followers that the change is committed.

Let's take a look at an example of how changes go through the leader and are replicated to all the nodes in a Raft group. In this example, there are three nodes in a Raft group. **Node 2** is the leader, and **Node 1** and **Node 3** are followers, as shown in the following diagram:

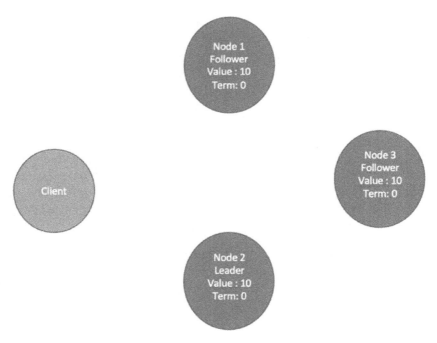

Figure 2.6 – Raft group with Node 2 as the leader and node value set to 10

Let's look at the changes through the following steps:

1. Now, a client makes a request to the Raft leader to change the value from **10** to **30**, as illustrated in the following screenshot:

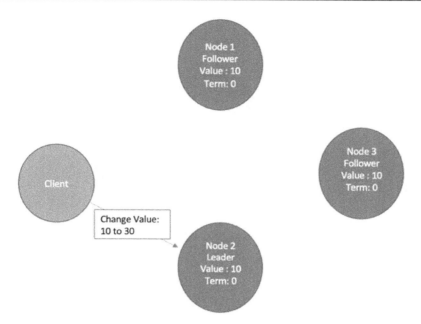

Figure 2.7 – Client requesting the node leader to change the value from 10 to 30

2. When the leader receives this request, it appends this new entry of changing the value from **10** to **30** in its replication log, as illustrated in the following screenshot:

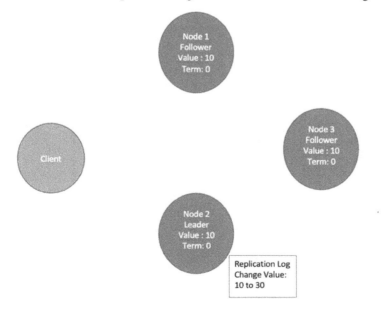

Figure 2.8 – Node 2 appends the entry to change the value from 10 to 30 in its replication log

3. After making an entry in its replication log, it is broadcast to all its followers, as illustrated in the following screenshot:

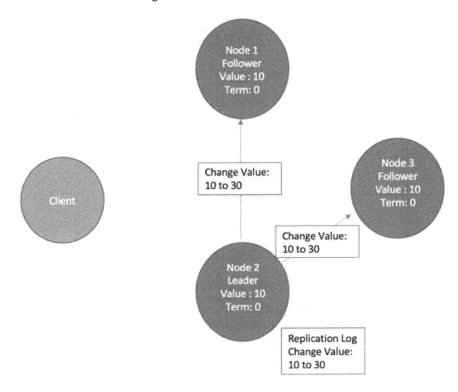

Figure 2.9 – Node 2, the leader, broadcasts this entry of changing the value from 10 to 30 to all its followers

4. All the followers now make an entry in their replication log and acknowledge to the leader that they have done so, as illustrated in the following screenshot:

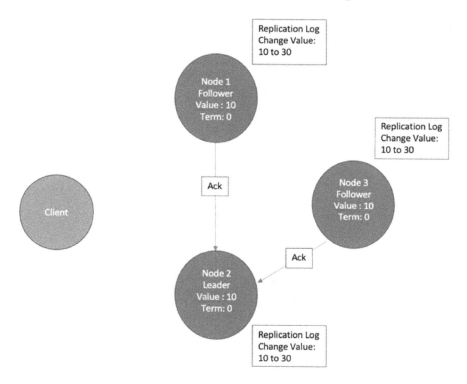

Figure 2.10 – Followers acknowledge appending the new entry in their own replication logs

5. Once the leader receives an acknowledgment from a majority of the followers, it commits this entry locally and changes the actual node value from 10 to 30, and later broadcasts to all its followers that the entry is committed. This is illustrated in the following screenshot:

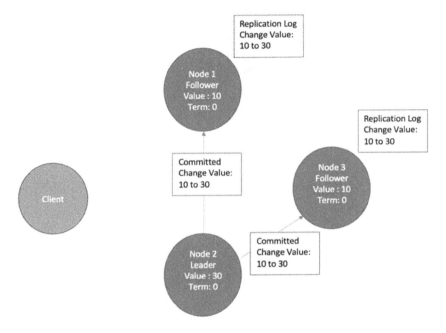

Figure 2.11 – Once a majority of the followers acknowledge, the leader commits the changes locally and sends a notification to all its followers about the commit.

6. In the last step, all the followers also commit the entry by changing their node values from **10** to **30**, as illustrated in the following screenshot:

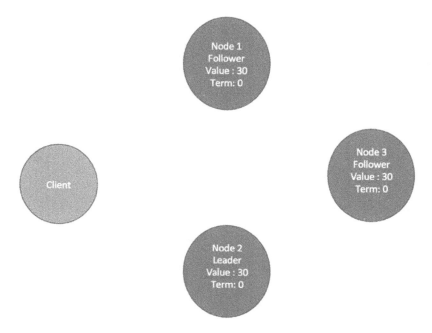

Figure 2.12 – All the followers acknowledge appending the new entry in their own replication logs

This completes the process of log replication for a single change. This process is replayed for all the changes.

If the leader crashes during this negotiation, the replication log can be in an inconsistent state. The new leader fixes this inconsistency by forcing its followers to replicate its own log. Specific details on how this is done are beyond the scope of this book.

Next, we will understand how CockroachDB interacts with the disk.

Interactions with the disk for data storage

The storage layer is responsible for reading from and writing to the disk. Each node in a CockroachDB cluster should have at least one storage attachment. Data is stored as key-value pairs on the disk using a storage engine.

Storage engine

A **database management system (DBMS)** uses a storage engine to perform **CRUD** (which stands for **create**, **read**, **update**, and **delete**) operations on the disk. Usually, the storage engine acts as a black box, so you get more options to choose from based on your own requirements, and also, storage engines evolve independently of the DBMSes that are using them.

Storage engines use a variety of data structures to store data. The popular ones are listed here:

- **Hash table**
- **B+ tree**
- **Heap**
- **Log-structured merge-tree (LSM-tree)**

Storage engines also work with a wide range of storage types, including the following:

- **Solid-state drive (SSD)**
- Flash storage
- Hard disk
- Remote storage

CockroachDB primarily supports Pebble as the storage engine, as of version 21.1. Previously, it also supported RocksDB.

Let's look at Pebble.

Pebble

Pebble is primarily a key-value store that provides atomic write batches and snapshots. Starting from version 20.2, CockroachDB uses the Pebble storage engine by default. Pebble was developed to address two concerns, which are outlined here:

- Focusing the storage engine's features to primarily address the requirements of CockroachDB.
- Improving the performance by bringing in certain optimizations that are not part of RocksDB.

Also, since it's developed by Cockroach Labs engineers, it's easy to maintain and control its roadmap. This has also increased the overall productivity as Pebble is written in **Golang**, just as is CockroachDB itself.

Summary

In this chapter, we learned about all the layers of CockroachDB and how a given request is processed through these layers. We also went through how queries are handled; how a transactional key-value store works; Raft, a distributed consensus algorithm; and a bit about storage engines.

In the next chapter, we will understand ACID and how it's implemented in CockroachDB.

Section 2: Exploring the Important Features of CockroachDB

In this section, we will go over several important features of CockroachDB. This should give you a better insight with respect to when to use CockroachDB as a datastore.

This section comprises the following chapters:

- *Chapter 3, Atomicity, Consistency, Isolation, and Durability (ACID)*
- *Chapter 4, Geo-Partitioning*
- *Chapter 5, Fault Tolerance and Auto-Rebalancing*
- *Chapter 6, How Indexes Work in CockroachDB*

3
Atomicity, Consistency, Isolation, and Durability (ACID)

In the *Chapter 2, How Does CockroachDB Work Internally*, we learned about the different layers of **CockroachDB**. In this chapter, we will learn about what **ACID** is, its importance, and what the ACID guarantees that CockroachDB provides are.

ACID guarantees the following things:

- **Atomicity**: This is achieved through the notion of a transaction, in which all the statements within a transaction are executed as a single unit. So, either all of them succeed or fail together.

- **Consistency**: The database state should be consistent before and after a transaction is executed and should ensure that the database constraints are never violated.

- **Isolation**: Multiple transactions can get executed independently at the same time, without running into each other.
- **Durability**: Changes, once committed, always remain intact, irrespective of any system or network failures.

The following topics will be covered in this chapter:

- An overview of ACID properties
- ACID from CockroachDB's perspective

An overview of ACID properties

In this section, we will discuss each of the ACID properties and understand their importance in avoiding data loss and corruption. First, we will take a look at atomicity.

Atomicity

Atomicity refers to the integrity of a given transaction, which means if a transaction comprises multiple statements, atomicity ensures that either all of them succeed or none of them succeed. Atomicity is important to make sure that there is no data inconsistency because of a transaction getting partially executed.

Let's try to understand this with an example:

```
BEGIN TRANSACTION
Read Foo's Account
Debit $100 from foo's Account
Read Bar's Account
Credit $100 to bar's Account
COMMIT
```

Here, you have a transaction in which you are debiting the money from `Foo's Account` and crediting it to `bar's Account`. Here, it's important that these two activities happen as a single unit of work. Otherwise, it can result in data inconsistency. Consider a case in which you are able to debit the money from `Foo's Account`, but not able to credit it to `bar's Account` – you will lose track of the money that was deducted from `Foo's Account`. So, either all of them should succeed, or none of them should.

Next, we will learn about consistency.

Consistency

Consistency in ACID is an overloaded term and can mean several things, including the following:

- Ensuring that the transactions in the future see the effects of transactions that are already committed

- Ensuring database constraints are not violated once a given transaction is committed

- Ensuring that all the operations in a transaction are executed correctly

Basically, consistency is responsible for ensuring the database always moves from one valid state to another, which doesn't result in any data inconsistency or corruption.

In the context of a **Consistency, Availability, and Partition Tolerance (CAP)** theorem, consistency indicates that in a distributed system, all the reads receive the most recent write, or it will error out. As per the CAP theorem, you can only have two of consistency, partition tolerance, and availability. Since consistency will be important to most of the applications, you have to choose between availability and partition tolerance.

Next, we will go through various isolation levels and try to understand the implication of each on a transaction.

Isolation

Isolation deals with the guarantees a database provides when multiple clients are interacting with the same set of data.

Some of the popular isolation levels are as follows:

- **Serializable**: This is the highest isolation level which requires acquiring a lock on the data you are operating. Transactions in CockroachDB implement the highest isolation level which is serializable. This means that transactions will never result in inconsistent or corrupt data. In the case of CockroachDB, this is provided by using range-level locks called **write intents**.

- **Snapshot**: This is a non-lock-based concurrency control, so no locks are not used, however, if a conflict is detected between concurrent transactions, then only one of them is allowed to commit.

- **Read uncommitted**: This is the lowest isolation level which allows dirty reads, which means changes made by live transactions that are not yet committed.

- **Repeatable read**: Repeatable read guarantees that you only read a committed value and also already read data cannot be changed by some other transaction. However, it does allow a *phantom read*. A phantom read happens when, between two reads of the same data, some other parallel transaction adds new data and they show up in subsequent reads, once they are committed.

- **Read committed**: This isolation level guarantees that any read you do is already committed.

With an example, let's try to understand the difference between read committed, repeatable read, and serializable:

T1	T2
BEGIN TRANSACTION; SELECT * FROM FOO; SELECT SLEEP(10); SELECT * FROM FOO; COMMIT;	BEGIN TRANSACTION; SELECT SLEEP(5); DELETE 5 ROWS INTO TABLE FOO; INSERT 10 ROWS INTO TABLE FOO; COMMIT;

Figure 3.1 – Two transactions, T1 and T2, happening in parallel

In the previous example, there are two parallel transactions, T1 and T2.

For the discussion, let's assume that the following figure is the sequence of events:

1	T1.BEGIN TRANSACTION
2	T2.BEGIN TRANSACTION
3	T1. SELECT * FROM FOO;
4	T2.SELECT SLEEP(5);
5	T1.SELECT SLEEP(10);
6	T2.DELETE 5 ROWS FROM TABLE FOO;
7	T2.INSERT 10 ROWS INTO TABLE FOO;
8	T2.COMMIT;
9	T1.SELECT * FROM FOO;
10	T1.COMMIT;

Figure 3.2 – Ordering of operations for transactions T1 and T2

Here, serializable guarantees that the T1 transaction sees the exact same value in *Steps 3 and 8*, although there are new rows added and some rows got deleted by the T2 transaction.

With read committed, *Step 9* will see new rows added by the T2 transaction and it will not see the rows deleted by the T2 transaction, since all the changes by T2 are already committed.

In the case of a repeatable read, *Step 9* will only see the new data committed by T1.

Durability

Durability guarantees that any changes that are committed are permanent, irrespective of failures related to nodes, memory, storage, or the network. Databases achieve durability by flushing out the transactional log to non-volatile storage like solid-state drives and magnetic storage devices.

Many of the **database management systems** (**DBMS**) use the concept of a transaction log to ensure they can recover from system crashes. The transaction log is also called a binary log, database log, or audit trail. The transaction log is usually stored on an external storage device.

Let's say a database node crashes while executing a bunch of transactions. Now, whenever it comes back, it goes through the transaction log to determine which transactions were committed and which were uncommitted at the time of the crash. If a transaction is committed, all the changes made during that transaction are replayed. If a transaction was uncommitted, all the changes made by the transaction are rolled back.

Next, we will learn about how ACID properties are supported in CockroachDB.

ACID from CockroachDB's perspective

In this section, we will go over how each of the ACID properties is implemented in CockroachDB and what guarantees they provides. Like in the previous section, we will start with atomicity.

Atomicity

As we learned in the first section, atomicity ensures that all the statements in a given transaction are executed as a single unit – that is, either all of them succeed or all of them fail. This condition should be guaranteed irrespective of machine, network, and memory failures. This is essential to make sure multiple queries don't run into each other.

CockroachDB allows you to have ACID transactions that can span the entire cluster, touching multiple nodes and geographical locations. CockroachDB supports this using an atomic protocol called **parallel commits**.

In the previous chapter, we learned about *transaction records* and *write intent*. A transaction record keeps track of the current state of the transaction and is maintained in the range where the first write occurs. Whenever we are changing a value, they are not directly written to the storage layer. Instead, a value is written to an intermediate state known as `write intent`. Write intent acts on a **multiversion concurrency control (MVCC)** record, with a link to the transaction record. Write intent acts as a replicated lock, which houses a replicated provisional value.

Write intent has been shown in the following figure with a sample transaction with two writes:

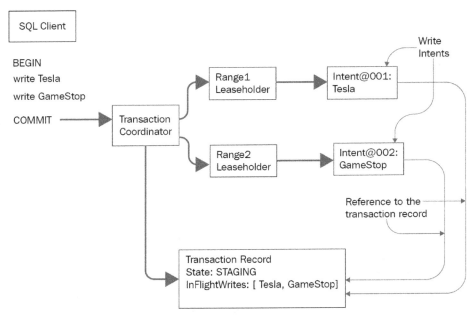

Figure 3.3 – Transaction timeline without parallel commit

Any transaction that comes across a write intent should also go through corresponding transaction records and, based on its state, decide how to treat the write intent. A commit flips the transaction record state to `committed`. Once the transaction is committed, write intents are cleaned up asynchronously.

Now, let's take a look at how traditional atomic transactions worked in CockroachDB without parallel commits and later with parallel commits.

Atomic transactions without parallel commits

Prior to version *19.2*, there was no parallel commit concept and transactions worked similar to a two-phase commit protocol. Let's explore this with an example.

Let's say there is a transaction with three writes, as shown in the following example:

```
BEGIN
write Tesla
write GameStop
write Amazon
COMMIT
```

This entire flow has been depicted in *Figure 3.4*:

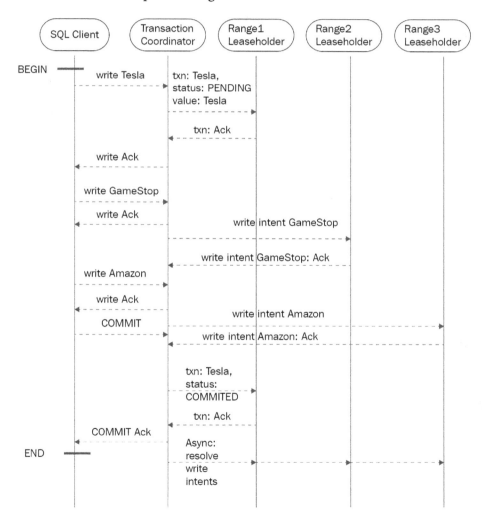

Figure 3.4 – Transaction timeline without parallel commit

Consider the following:

1. The transaction record is created in the first range where the write happens. In this case, that happens to be the Tesla value. So, along with the provisional Tesla value, a write intent is created with the transactional record in a PENDING state.

2. As and when the rest of the writes are received, they create their own write intents. All these write intents are said to be pipelined because CockroachDB doesn't wait until they succeed before receiving the next statement within the same transaction from the **SQL** client.

3. Once the COMMIT is issued in the transaction, CockroachDB waits for all the write intents to succeed with replication.

4. Once all of them succeed, the transaction record state is changed to COMMITTED. Again, this is replicated for durability and the transaction is considered committed once the replication is completed.

5. At this time, CockroachDB sends back an acknowledgment that the transaction is committed.

6. Later, the rest of the write intents are resolved and are eventually cleaned up.

This mechanism is similar to a two-phase commit protocol, in which write intents can be compared to the prepare phase and marking the transaction record as committed is similar to the commit phase. The main problem here is that it is blocking, since CockroachDB waits for all the write intents to succeed before marking the transaction record as committed. If the coordinator node crashes, then it's impossible to recover from that.

CockroachDB overcomes this problem by implementing a two-phase commit on top of a consensus protocol, **RAFT**, which we discussed in the previous chapter. This ensures that the transaction records themselves are replicated and highly available, and can recover from a coordinator crash.

Having a two-phase commit protocol on top of RAFT introduces more latency. This is because CockroachDB first waits for all the write intents to succeed before changing the transaction status to committed. Later, it has to wait until changing the transaction status itself has succeeded, as that involves one more round of consensus.

Now, let's see how an atomic transaction with parallel commit avoids this added transactional latency.

Atomic transactions with parallel commits

Parallel commit was introduced to reduce the transaction latency observed in the previously discussed two-phase commit like protocol

In the previous protocol, the transaction record has to wait until all the write intents have succeeded to change the status to committed. In parallel commit, there is a new status called STAGING. The transaction record also includes the list of all the keys for which there are write operations in the current transaction. A transaction can be implicitly assumed to be committed if all the writes that are listed in the transaction record have succeeded and reached consensus.

Let's go over the same transaction with three writes:

```
BEGIN
write Tesla
write GameStop
write Amazon
COMMIT
```

This entire flow has been depicted in *Figure 3.5*:

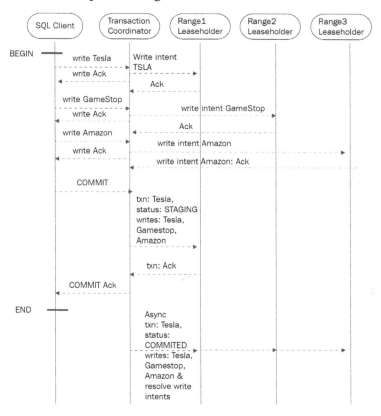

Figure 3.5 – Transaction timeline with parallel commit

Here, the key difference is that the logic of a transaction being committed depends on the status of all the writes involved in that transaction. Also, if we know the status of all the writes, we don't have to wait until the transaction record status is explicitly updated to committed after reaching consensus. Because of this, the transaction's coordinator node can acknowledge that a given transaction has been committed successfully to the SQL client once the coordinator observes that all the writes in that transaction have succeeded. The other important change in this protocol is that the initiation of pipelining of the write to the transaction record with the STAGING status is done after a COMMIT for a given transaction is received from the SQL client. Pipelining the write to the transaction record is done in parallel with pipelining the write intents, in order to speed up the entire process.

Now, let's take a look at how a transaction status is recovered whenever a transaction coordinator crashes in the middle of a transaction.

Transaction status recovery

Now, let's see what happens if the coordinator crashes before it can update the transaction record to either COMMITTED or ABORTED. In this case, whenever there is a transaction with a conflicting write intent, it looks up that write intent's transaction record. Once it sees that the status is STAGING, it cannot decide whether that transaction was COMMITTED or ABORTED. So, now it starts the status recovery process.

During transaction status recovery, each write intent involved in that transaction is consulted to see if it succeeded. If all the write intents have succeeded, the transaction is assumed to be COMMITTED, and if not, to be ABORTED. After this, the appropriate status is updated so that any other conflicting transaction in the future doesn't have to go through the status recovery process again.

Status recovery can be very expensive, especially if it involves multiple writes with ranges that do not share leaseholders. If multiple leaseholders are involved in status recovery, there will be multiple roundtrips to several nodes, before we can recover the status. To avoid this, CockroachDB does two things:

1. The transaction coordinator node marks the transaction record as COMMITTED or ABORTED as soon as it can.

2. Transaction coordinators periodically send heartbeats to their transaction records. This helps the conflicting transactions to determine if a transaction is still alive or not.

In the next section, we will learn about how consistency is ensured within CockroachDB.

Consistency

As discussed earlier, consistency deals with two things:

1. Ensuring no database rules are violated
2. Making sure that transactions that are executed in parallel on the same set of data do not conflict with each other, which is necessary to avoid data consistency issues

For the first one, it boils down to making sure that the database doesn't have any bugs and does whatever it claims. **Jepsen** is an effort to improve the safety of distributed databases, queues, and consensus systems. During Jepsen testing, a given system is verified for whether it lives up to its documentation's claims. CockroachDB passed Jepsen testing in 2017.

In CAP theorem, which we discussed in the first chapter, consistency means every read sees the latest write or errors out. CockroachDB is a **consistent and partition tolerant (CP)** system, which means its highly consistent and, whenever there are partitions, the system becomes unavailable rather than ending up with inconsistent data.

Let's now learn about isolation and what kind of isolation CockroachDB provides.

Isolation

CockroachDB uses something called a **serializable snapshot**, which is an optimistic, multi-version, timestamp-ordered concurrency control system.

It's a distributed, lockless, recoverable, and serializable protocol. Distributed, as multiple nodes can be involved. Lockless, as operations are performed without locks and correctness is ensured by aborting transactions that violate serializability. Recoverable, since aborted transactions don't have any effect on the state of the database, which is ensured by the atomic commit protocol. Serializable, since CockroachDB guarantees a consistent database state by ensuring serial execution of composite transactions is correct.

Next, we will learn about durability in CockroachDB.

Durability

Durability guarantees that any changes that are committed are permanent. CockroachDB uses the RAFT consensus algorithm to ensure that all writes for a transaction record and write intents are durable. We have already discussed RAFT at length in *Chapter 2, How Does CockroachDB Work Internally?*

CockroachDB replicates each range three times by default and ensures that each replica is stored on different nodes. If a minority of the nodes fail, CockroachDB continues to operate and does not result in inconsistency or loss of data.

Let's take a look at how durability works in a three-node cluster:

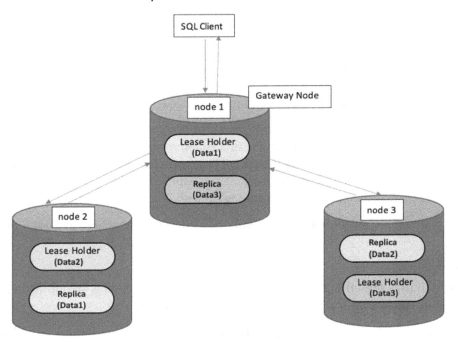

Figure 3.6 – Transaction timeline with parallel commit

As you can see, **node 1** is the lease holder for **data1** and **node 2** has the replica. Similarly, **node 3** is the lease holder for **data3**, and **node 1** has the replica. Next, **node 2** is the lease holder for **data2**, and **node 3** has the replica. Here, **node 1** is also acting as a gateway node, where the initial request from the SQL client lands, and it also coordinates other nodes in the cluster to serve the request:

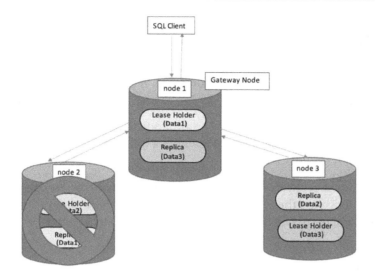

Figure 3.7 – Transaction timeline with parallel commit

Now, let's say **node 2** is not available due to a system failure. Since **node 2** was the lease holder of **data2**, now the coordinator is unable to get the data for **data2** from **node 2**, since it's not available. Now, the RAFT group for **data2** will hold an election and the lease holder will be reassigned. In this case, it has to be **node 3**, as that's the only other node that has the replica of **data2**:

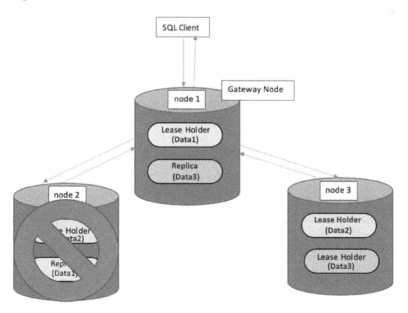

Figure 3.8 – Transaction timeline with parallel commit

As you can see in *Figure 3.8*, **node 3** acts as the lease holder for both **data2** and **data3** and the cluster is still fully functional, in spite of a node failure.

Now, if **node 3** also goes down for some reason, since we don't have any lease holder for **data2**, if a request involves serving the data for **data2**, it cannot be completed, affecting the availability of **data2** on a given CockroachDB cluster.

As you can see, CockroachDB can tolerate certain node failures, as long as there are lease holders for all the data. Otherwise, availability takes a hit. In either case, there won't be any data loss.

Users can also configure replication zones for databases, tables, rows, indexes, and system data. We will discuss these configurations in *Chapter 5, Fault Tolerance and Auto-Rebalancing*, where we will discuss fault tolerance and auto-recovery.

Summary

In this chapter, we learned about the four basic pillars of any database: **Atomicity, Consistency, Isolation, and Durability**. As a recap, atomicity ensures that a transaction is executed as a single unit of work. Consistency involves making sure that any database operation doesn't violate any of the database constraints. In the context of a CAP theorem, consistency refers to the fact you will never read stale or uncommitted data. CockroachDB provides both serializable and snapshot isolation levels. CockroachDB uses the RAFT protocol for transaction records and write intents to guarantee that all committed data is durable and permanent, irrespective of node failures.

In the next chapter, we will go over fault tolerance and auto-recovery, and what some of the configurations are in CockroachDB.

4

Geo-Partitioning

In the *Chapter 3, Atomicity, Consistency, Isolation, and Durability (ACID)*, we learned about what ACID is, why we need it, and how it's supported in CockroachDB. Here, we will learn all about geo-partitioning. Geo-partitioning is one of the most important reasons why you will want to use a distributed SQL database such as CockroachDB.

In this chapter, you will get a basic understanding of what geo-partitioning is and why this feature is useful for you. We will also go over some cloud jargon and some of the options provided by various cloud providers to distribute your data geographically for better resiliency, performance, and availability. At the end of the chapter, we will go over different ways of geo-partitioning your data in CockroachDB.

The following topics will be covered in this chapter:

- Introduction to geo-partitioning
- Cloud regions and zones
- Geo-partitioning in CockroachDB

Technical requirements

For executing the examples in this chapter, you will need to install CockroachDB. If you still haven't done so, please refer to the *Technical requirements* section in *Chapter 2, How Does CockroachDB Work Internally?* All the queries in this chapter are available at `https://github.com/PacktPublishing/Getting-Started-with-CockroachDB`.

Introduction to geo-partitioning

As the word *geo-partition* suggests, the data is partitioned based on geographical locations. Geo-partitioning refers to the mechanism of storing the data in various geographical locations, based on where the data is being consumed.

For example, let's say you are maintaining a database for an airlines company that has international and domestic travelers as its users from every continent. Since they have a global presence, it would be beneficial to keep the users' data close to where they live. This will help in serving the data locally and quickly.

Figure 4.1 shows an example of a table whose rows are partitioned based on geo-location across three different continents. Rows are stored in specific databases based on their locality. This locality can be mapped to the user's location based on their activity:

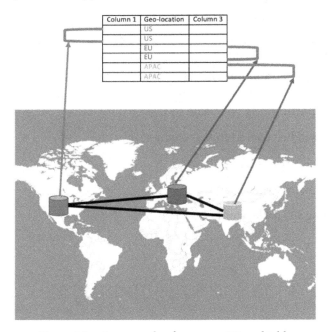

Figure 4.1 – An example of a geo-partitioned table

Geo-partitioning will be useful, even within smaller regions, where you provide services locally. An example would be, let's say, you are managing a database for a company that has to serve lots of data very frequently, which is highly localized to users living in that area. This example fits companies such as Instacart, DoorDash, Uber Eats, Uber, Lyft, and Ola.

For the sake of discussion, let's consider DoorDash. In the case of DoorDash, its users search for restaurants and grocery stores, usually within 10–15 miles of where they live, and place an order on their app. In this case, we can house the data regarding user purchase orders and delivery status close to the users. This not only helps us in serving the data faster for local users, but we can also easily apply local compliance rules only on a specific subset of CockroachDB clusters.

If we consider the state of California for this example, we can strategically deploy CockroachDB nodes in San Francisco and Los Angeles. For restaurants, menus, and grocery items, which don't change often, we can create duplicate indexes that will help reduce read latencies when browsing.

Also, now, if you want to incorporate the **California Consumer Privacy Act of 2018 (CCPA)** for all the purchase orders that happen within California, it would be much easier to restrict it to the databases in California that manage this information. The following figure illustrates this aspect:

Figure 4.2 – Geo-partitioning for cities in California State, USA

Some of the main advantages of geo-partitioning are as follows:

- It helps serve data quickly, as we reduce the number of network hops.
- Failures in a given geolocation only affect a small set of users.
- It provides data compliance as per local rules and standards, for example, the **California Consumer Privacy Act of 2018 (CCPA)** and the European Union's **General Data Protection Regulation (GDPR)**.
- It helps provide better disaster recovery and resiliency if an entire geolocation experiences a natural or human-triggered calamity.

Let's now learn about cloud regions and availability zones, which is essential to understand how geo-partitioning can be realized on various cloud providers.

Cloud, regions, and zones

In this section, we will learn about some jargon related to the cloud. We will also learn about regions and zones, and how they have been realized by various cloud providers. These concepts are important to understand to decide how you want to distribute your data and what sort of guarantees you want to provide:

- **Cloud**: A cloud is nothing but a bunch of servers on multiple data centers that are positioned in strategic locations across the globe. These data centers provide resources such as storage, network, and compute on demand, and they belong to a specific cloud provider.
- **Cloud provider**: A cloud provider is an organization that provides various services on its public or private cloud platform.
- **Public cloud**: In a public cloud, resources that you consume are hosted on the cloud provider's data center. The cloud provider is responsible for maintaining, upgrading, and operating cloud resources. Since you are consuming resources maintained by a third party, there are additional security risks here.
- **Private cloud**: In the case of the private cloud, the resources are usually hosted on a company's own data center, but they can also be hosted by a cloud provider. In a private cloud, all the resources are dedicated to a single organization and isolated from other organizations; hence it's more controlled and secured.
- **Multi-cloud**: A given platform is called multi-cloud where you consume resources from multiple public cloud providers.
- **Hybrid cloud**: In a hybrid cloud, you will be combining resources from a public cloud along with resources from a private cloud and/or on-premises.

Region

A **region** refers to an actual physical location where your cloud resources are housed. Each cloud provider has different notions of a region.

It's very important to understand how regions are implemented by different cloud providers, as it determines the following things:

- **Cloud cost**: Resources in some regions are cheaper than in others.
- **Multi-cloud and hybrid cloud strategy**: This includes disaster recovery, high availability, data replication, data migration, data sharing, failover, and so on.
- **Latency**: The whole idea of geo-partitioning is about reducing the latency by keeping the data close to the customer. So, it becomes apparent that there is a need to select a region in strategic locations, which reduces overall latency.
- **Data compliance**: Depending on where the region is located, you might have different data compliance requirements. Also, some countries might insist that the data of their citizens cannot leave the country, in which case you will be forced to pick some regions in that country.
- **Services and features**: Not all services and features are available in all regions. So, this sometimes reduces the choice of regions.

Zone

A region consists of multiple zones. A **zone** refers to a more specific location within a given region.

Availability zone

An **availability zone** is an isolated data center that doesn't share any resources with other zones within the same region. All the communication between availability zones happens through a high-speed network. A region is supposed to have at least two availability zones that help in implementing redundancy, failover, and high availability.

Now, let's understand the definitions of region and zone by some of the top cloud providers.

Regions and zones on various cloud providers

In this section, we will briefly go over what region and zone mean on the top four cloud providers. We will be covering the following cloud providers in this section:

- Amazon Web Services
- Google Cloud Platform
- Microsoft Azure
- Oracle Cloud

Let's get started!

Amazon Web Services

Region: A Region is a physical location that consists of multiple data centers.

Availability Zone: A group of discrete data centers that provide redundancy to cloud resources is called an Availability Zone. Availability Zones help in implementing features such as high availability, fault tolerance, reliability, and scale.

AWS Local Zone: Local Zones provide resources that are located close to your end users. These will be useful in services such as gaming and streaming, which require low latency, high throughput, and elastic services.

Google Cloud Platform

Region: A region is a collection of zones.

Zones: A zone is a deployment within a region. You should use multiple zones to provide high availability and fault tolerance.

Microsoft Azure

Region: A region is a set of data centers connected within a perimeter determined by the latency and connected through a fast network.

Geography: An area of the world containing at least one Azure region. A geography spans multiple regions and is fault-tolerant, even in the event of a complete regional failure.

Availability zones: Unique physical locations within a region. Each availability zone comprises one or more data centers with resources that are not shared with other zones.

Oracle Cloud

Region: A region is a localized area and is made of several availability domains.

Availability domains: Availability domains are made of one or more data centers, they do not share any resources amongst them, and are connected through a fast network.

Each availability domain has three **fault domains**. Fault domains ensure your resources are from different availability domains, which offers improved resiliency.

Next, we will learn about how to achieve geo-partitioning with CockroachDB.

Geo-partitioning in CockroachDB

CockroachDB provides two topology patterns, which provide two levels of data resiliency, latency and availability.

Single region

Here, the entire data is in a single region.

CockroachDB defines two variations of single-region topology, development and production, as follows:

- **Development**: This pattern is very straightforward, where you just have a single node in an availability zone, with multiple clients talking to it. This pattern is useful for testing purposes. This topology can also be used on your laptop or desktop.

  As part of your **Continuous Integration/Continuous Deployment (CI/CD)** pipeline, you can have a dedicated stage in which you provision a single-node cluster and later can run a bunch of system tests that interact with a real database. Since the clients will be local to the data, reads and writes will be much faster, although there is no resiliency.

The following is an example of a single-region deployment:

Figure 4.3 – Single-node deployment in the US-West (Northern California) region, where all the clients are also deployed in the same region

- **Basic production**: Here, you can have nodes deployed in more than one availability zone within the same region. It is ideal to have at least three nodes in three different availability zones within the same region for consensus purposes. This pattern takes advantage of many CockroachDB features, such as **replication**, **rebalancing**, and **resiliency**.

This topology can withstand up to a single-node failure. If two nodes fail, then some ranges might not have any leaseholders due to a lack of consensus and will become unavailable. You would also need a load balancer to spread the traffic from clients across three nodes evenly.

The following is a single-region deployment with three nodes:

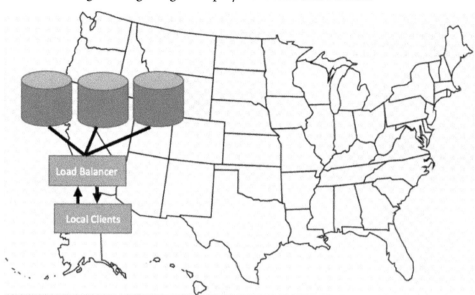

Figure 4.4 – Single-region three-node deployment in the US-West (Northern California) region, where all the clients are also deployed in the same region

In basic production topology, client requests hit a load balancer first, and later they are forwarded to one of the nodes, which acts as a gateway node. The gateway node then coordinates with relevant leaseholders, gathers all the data, and serves the data back to the client.

Multi-region

Here, the data is spread across multiple regions. You can use row-level control to distribute the rows geographically. The following figure shows a multi-region deployment, which spans three different regions in the US:

Figure 4.5 – Multi-region nine-node deployment covering the US-East, US-Central, and US-West regions

You can choose the following topologies based on your requirement:

- Geo-partitioned replicas
- Geo-partitioned leaseholders
- Duplicate indexes
- Follower reads
- Follow-the-workload

Geo-partitioning options are only available with the enterprise license of CockroachDB. If you try to use enterprise features without the enterprise license, you will see the following message:

```
ERROR: use of partitions requires an enterprise license. see
https://cockroachlabs.com/pricing?cluster=65244c3a-2d63-432c-
a8b4-c70a53459cal for details on how to enable enterprise
features
SQLSTATE: XXC02
```

You can visit the URL mentioned in the error message and get an enterprise license for a specific trial period.

Applying an enterprise license to your cluster would involve the following steps:

```
root@localhost:26258/defaultdb> SET CLUSTER SETTING cluster.
organization = 'Self';
SET CLUSTER SETTING


Time: 166ms total (execution 165ms / network 0ms)


root@localhost:26258/defaultdb> SET CLUSTER SETTING enterprise.
license = 'crl-0-EKD0mYsGGAIiBFN1bGY';
SET CLUSTER SETTING


Time: 191ms total (execution 191ms / network 0ms)
```

Now, let's take a deeper look at each one of these topologies. All these topologies would need a multi-region deployment. The easiest way to create a multi-region cluster for experiment purposes is to create an account at `https://cockroachlabs.cloud` and request a three-region, nine-node cluster.

Geo-partitioned replicas

In the case of geo-partitioned replicas, you have to have a column for the geolocation. It then has to be combined with the table's unique identifier to form a compound primary key. For example, let's say you have an ID, which is a **Universally Unique Identifier (UUID)** – you can combine that with the city or the state to form a compound primary key. Then, you have to partition the table and all the secondary indexes based on that column, and each partition will have its own replicas. Once you have this, you can ask CockroachDB to place data belonging to each partition in a specific region.

For example, let's say you have users from California and Ohio. All the rows belonging to users from California can be stored in the US-West region and users in Ohio can be stored in the US-East region. Here, the assumption is that data will be consumed locally, so both read and write latencies will be reduced.

For example, if you operate a food delivery service, your users will place orders from the city they live in, and also the food will be delivered within the same city. So, in this case, it makes total sense to use a geo-partitioned replica for your users. Since all the data and its replica are housed in the same region, if an entire region goes down, the data will not be available.

Let's see an example in which we are going to create a geo-partitioned replicas topology. Basically, the replicas will be pinned to a particular region so that local reads and writes are faster:

1. Create a table called `users`, where you have `city` as one of the columns. This will help in partitioning this table by city:

    ```
    CREATE TABLE users (
        id UUID NOT NULL DEFAULT gen_random_uuid(),
        first_name STRING NOT NULL,
        last_name STRING NOT NULL,
        city STRING NOT NULL,
        PRIMARY KEY (city ASC, id ASC) );
    ```

2. Now, create a secondary index as follows:

    ```
    CREATE INDEX users_first_name_last_name_index ON users
    (city, first_name, last_name);
    ```

3. Create partitions for the table based on `city`. Let's consider three different cities that are in the west, east, and central parts of the USA:

    ```
    ALTER TABLE users PARTITION BY LIST (city) (
        PARTITION sfo VALUES IN ('san francisco'),
        PARTITION aus VALUES IN ('austin'),
        PARTITION ny VALUES IN ('new york')
    );
    ```

4. Create partitions for the secondary index based on `city`:

```
ALTER INDEX users_first_name_last_name_index PARTITION BY
LIST (city) (s
    PARTITION sfo VALUES IN ('san francisco'),
    PARTITION aus VALUES IN ('austin'),
    PARTITION ny VALUES IN ('new york')
);
```

5. For the table and its secondary index, create a replication zone, which will pin the replica of a given partition to its relevant region. The `<table>@*` syntax lets you create zone configurations for all identically named partitions of a table, saving you multiple steps:

```
ALTER PARTITION sfo OF INDEX users@* CONFIGURE ZONE USING
constraints = '{"+region=us-west":1}', num_replicas=3;


ALTER PARTITION aus OF INDEX users@* CONFIGURE ZONE USING
constraints = '{"+region=us-central":1}', num_replicas=3;


ALTER PARTITION ny OF INDEX users@* CONFIGURE ZONE USING
constraints = '{"+region=us-east":1}', num_replicas=3;
```

6. Now, let's execute the `SHOW CREATE TABLE` query to see how the partitions are created for the table and the secondary index:

```
SHOW CREATE TABLE users;
```

The sample output is as follows:

```
SHOW CREATE TABLE users;
   table_name
|                                                      create_
statement
--------------+--------------------------------------------------
-------------------------------------------------------------
--------------
   users      | CREATE TABLE public.users (
             |     id UUID NOT NULL DEFAULT gen_random_
uuid(),
             |     first_name STRING NOT NULL,
             |     last_name STRING NOT NULL,
```

```
|           city STRING NOT NULL,
|           CONSTRAINT "primary" PRIMARY KEY (city
ASC, id ASC),
|           INDEX users_first_name_last_name_index
(city ASC, first_name ASC, last_name ASC) PARTITION BY
LIST (city) (
|             PARTITION sfo VALUES IN (('san
francisco')),
|             PARTITION aus VALUES IN
(('austin')),
|             PARTITION ny VALUES IN (('new
york'))
|           ),
|           FAMILY "primary" (id, first_name,
last_name, city)
|         ) PARTITION BY LIST (city) (
|           PARTITION sfo VALUES IN (('san
francisco')),
|           PARTITION aus VALUES IN (('austin')),
|           PARTITION ny VALUES IN (('new york'))
|         );
|     ALTER TABLE defaultdb.public.users
CONFIGURE ZONE USING
|             num_replicas = 3;
|     ALTER PARTITION sfo OF INDEX defaultdb.
public.users@primary CONFIGURE ZONE USING
|             num_replicas = 3,
|             constraints = '{+region=us-west: 1}';
|     ALTER PARTITION sfo OF INDEX defaultdb.
public.users@users_first_name_last_name_index CONFIGURE
ZONE USING
|             num_replicas = 3,
|             constraints = '{+region=us-west: 1}';
|     ALTER PARTITION aus OF INDEX defaultdb.
public.users@primary CONFIGURE ZONE USING
|             num_replicas = 3,
|             constraints = '{+region=us-central:
1}';
|     ALTER PARTITION aus OF INDEX defaultdb.
```

```
public.users@users_first_name_last_name_index CONFIGURE
ZONE USING
              |            num_replicas = 3,
              |            constraints = '{+region=us-central:
1}';
              | ALTER PARTITION ny OF INDEX defaultdb.
public.users@primary CONFIGURE ZONE USING
              |            num_replicas = 3,
              |            constraints = '{+region=us-east: 1}';
              | ALTER PARTITION ny OF INDEX defaultdb.
public.users@users_first_name_last_name_index CONFIGURE
ZONE USING
              |            num_replicas = 3,
              |            constraints = '{+region=us-east: 1}'
(1 row)


Time: 278ms total (execution 278ms / network 0ms)
```

Here, you can see that replicas are constrained to relevant regions. So, if an entire region goes down, a partition in that region becomes unavailable.

Next, we will go through the geo-partitioned leaseholders topology.

Geo-partitioned leaseholders

Like in the case of a geo-partitioned replica, you still need a column that has geolocation. You will also need a compound primary key, which is a combination of a unique ID and geolocation.

Here, the main difference is that you only pin the leaseholder to a specific location, but the replicas can be stored in different regions. Since we are only pinning the leaseholder, reads will always be faster but writes take more time, since data is replicated across regions, which takes more time as replication also involves consensus.

Let's see an example, in which we are going to create a geo-partitioned leaseholders topology. Basically, the leaseholder will be pinned to a particular region, so that local reads are faster:

1. Create a table called `users`, where you have `city` as one of the columns. This will help in partitioning this table by city:

```
CREATE TABLE users (
    id UUID NOT NULL DEFAULT gen_random_uuid(),
    first_name STRING NOT NULL,
    last_name STRING NOT NULL,
    city STRING NOT NULL,
    PRIMARY KEY (city ASC, id ASC) );
```

2. Now, create a secondary index as follows:

```
CREATE INDEX users_first_name_last_name_index ON users
(city, first_name, last_name);
```

3. Create partitions for the table based on `city`. Let's consider three different cities that are in the west, east, and central parts of the USA:

```
ALTER TABLE users PARTITION BY LIST (city) (
    PARTITION sfo VALUES IN ('san francisco'),
    PARTITION aus VALUES IN ('austin'),
    PARTITION ny VALUES IN ('new york')
);
```

4. Create partitions for the secondary index based on `city`:

```
ALTER INDEX users_first_name_last_name_index PARTITION BY
LIST (city) (
    PARTITION sfo VALUES IN ('san francisco'),
    PARTITION aus VALUES IN ('austin'),
    PARTITION ny VALUES IN ('new york')
);
```

5. For the table and its secondary index, create a replication zone, which will pin the leaseholder of a given partition to its relevant region:

```
ALTER PARTITION sfo OF INDEX users@*
    CONFIGURE ZONE USING
        num_replicas = 3,
        constraints = '{"+region=us-west":1}',
        lease_preferences = '[[+region=us-west]]';

ALTER PARTITION aus OF INDEX users@*
    CONFIGURE ZONE USING
        num_replicas = 3,
        constraints = '{"+region=us-central":1}',
        lease_preferences = '[[+region=us-central]]';

ALTER PARTITION ny OF INDEX users@*
    CONFIGURE ZONE USING
        num_replicas = 3,
        constraints = '{"+region=us-east":1}',
        lease_preferences = '[[+region=us-east]]';
```

6. Now, let's execute SHOW CREATE TABLE to see how the partitions are created for the table and the secondary index:

```
SHOW CREATE TABLE users;
```

The sample output is as follows:

```
SHOW CREATE TABLE users;
  table_name
|                                                           create_
statement
-------------+----------------------------------------------------
-------------------------------------------------------------------
--------------
  users      | CREATE TABLE public.users (
             |     id UUID NOT NULL DEFAULT gen_random_
uuid(),
```

```
    |         first_name STRING NOT NULL,
    |         last_name STRING NOT NULL,
    |         city STRING NOT NULL,
    |         CONSTRAINT "primary" PRIMARY KEY (city
ASC, id ASC),
    |         INDEX users_first_name_last_name_index
(city ASC, first_name ASC, last_name ASC) PARTITION BY
LIST (city) (
    |             PARTITION sfo VALUES IN (('san
francisco')),
    |             PARTITION aus VALUES IN
(('austin')),
    |             PARTITION ny VALUES IN (('new
york'))
    |         ),
    |         FAMILY "primary" (id, first_name,
last_name, city)
    | ) PARTITION BY LIST (city) (
    |         PARTITION sfo VALUES IN (('san
francisco')),
    |         PARTITION aus VALUES IN (('austin')),
    |         PARTITION ny VALUES IN (('new york'))
    | );
    | ALTER TABLE defaultdb.public.users
CONFIGURE ZONE USING
    |         num_replicas = 3;
    | ALTER PARTITION sfo OF INDEX defaultdb.
public.users@primary CONFIGURE ZONE USING
    |         num_replicas = 3,
    |         constraints = '{+region=us-west: 1}',
    |         lease_preferences = '[[+region=us-
west]]';
    | ALTER PARTITION sfo OF INDEX defaultdb.
public.users@users_first_name_last_name_index CONFIGURE
ZONE USING
    |         num_replicas = 3,
    |         constraints = '{+region=us-west: 1}',
    |         lease_preferences = '[[+region=us-
west]]';
```

```
          | ALTER PARTITION aus OF INDEX defaultdb.
public.users@primary CONFIGURE ZONE USING
          |          num_replicas = 3,
          |          constraints = '{+region=us-central:
1}',
          |          lease_preferences = '[[+region=us-
central]]';
          | ALTER PARTITION aus OF INDEX defaultdb.
public.users@users_first_name_last_name_index CONFIGURE
ZONE USING
          |          num_replicas = 3,
          |          constraints = '{+region=us-central:
1}',
          |          lease_preferences = '[[+region=us-
central]]';
          | ALTER PARTITION ny OF INDEX defaultdb.
public.users@primary CONFIGURE ZONE USING
          |          num_replicas = 3,
          |          constraints = '{+region=us-east: 1}',
          |          lease_preferences = '[[+region=us-
east]]';
          | ALTER PARTITION ny OF INDEX defaultdb.
public.users@users_first_name_last_name_index CONFIGURE
ZONE USING
          |          num_replicas = 3,
          |          constraints = '{+region=us-east: 1}',
          |          lease_preferences = '[[+region=us-
east]]'
  (1 row)


Time: 37ms total (execution 37ms / network 0ms)
```

Here, you can see that the lease preference is restricted to relevant regions, which will ensure that the leaseholders are always pinned to a specific region.

Next, we will go through the duplicate indexes topology.

Duplicate indexes

The duplicate indexes topology is useful in cases where you write once and read it from various locations. For example, let's say you are managing the credit cards of folks who travel throughout the US very often. If you pin the data to a single region, whenever the user moves out of that region, it will slow down the reads. So, duplicate indexes come in handy to solve this issue.

Just like the previous two cases, you will have a compound primary key with a combination of an ID and geolocation. Here, you can create a partition based on that column, but you only pin the leaseholder to a specific region.

In our example, it can be the primary address of the user. Here, the credit card information can be replicated in different regions to cover the entire US. Since only the leaseholder is responsible for writes and reads, your reads will always be routed to the pinned region of the leaseholder. This again introduces latency.

Now, you can create secondary indexes for the credit card. For example, assuming that your leaseholder is pinned to the west coast, you can create secondary indexes, such as `id_creditcard_east` and `id_creditcard_central`, which can be constrained to the US-East and US-Central regions respectively. This will also guarantee that there are local leaseholders for secondary indexes in all the regions, which drastically reduces the read latency, as they were served locally always.

Since we already have multiple copies of the original data and we are creating secondary indexes that are also replicated, now there are a lot of copies of the same data in multiple regions. So, this increases the write latencies, as all these copies have to be updated and a multi-region consensus has to be reached:

1. Let's say you are maintaining local attractions of the USA, which can be accessed by users throughout the USA:

    ```
    CREATE TABLE local_attractions (
        id UUID NOT NULL DEFAULT gen_random_uuid(),
        name STRING NOT NULL,
        address STRING NOT NULL,
        city STRING NOT NULL,
        PRIMARY KEY (id ASC)
    );
    ```

2. Create a replication zone and pin the leaseholder to a specific region:

```
ALTER TABLE local_attractions
    CONFIGURE ZONE USING
      num_replicas = 3,
      constraints = '{"+region=us-central":1}',
      lease_preferences = '[[+region=us-central]]';
```

3. Create secondary indexes for the other two regions. Here, storing a column improves the performance of queries that retrieve its values, but you cannot use these stored columns in the filtering logic:

```
CREATE INDEX idx_west ON local_attractions (city)
    STORING (name);
```

```
CREATE INDEX idx_east ON local_attractions (city)
    STORING (name);
```

4. For these secondary indexes, define the replication zone, once again pinning the leaseholder to the relevant region:

```
ALTER INDEX local_attractions@idx_west
    CONFIGURE ZONE USING
      num_replicas = 3,
      constraints = '{"+region=us-west":1}',
      lease_preferences = '[[+region=us-west]]';

ALTER INDEX local_attractions@idx_east
    CONFIGURE ZONE USING
      num_replicas = 3,
      constraints = '{"+region=us-east":1}',
      lease_preferences = '[[+region=us-east]]';
```

5. Now, let's execute SHOW CREATE TABLE to see how the partitions are created for the secondary indexes:

```
SHOW CREATE TABLE local_attractions;
```

The sample output is as follows:

```
SHOW CREATE TABLE local_attractions;
    table_name          |              create_statement
------------------------+-------------------------------------------
--------------
  local_attractions  | CREATE TABLE public.local_
attractions (
                       |     id UUID NOT NULL DEFAULT gen_
random_uuid(),
                       |     name STRING NOT NULL,
                       |     address STRING NOT NULL,
                       |     city STRING NOT NULL,
                       |     CONSTRAINT "primary" PRIMARY
KEY (id ASC),
                       |     INDEX idx_west (city ASC)
STORING (name),
                       |     INDEX idx_east (city ASC)
STORING (name),
                       |     FAMILY "primary" (id, name,
address, city)
                       | )
(1 row)

Time: 28ms total (execution 28ms / network 0ms)
```

Here, you can see that there are multiple identical indexes for multiple regions. So, whenever there are queries involving city and name, they can be served locally, hence reducing the latency of reads. Since we are maintaining identical indexes in multiple regions, the writes are much slower. This topology is useful where the data doesn't change much but is accessed frequently in all the regions.

Next, we will go through the follower reads topology.

Follower reads

If you want low-read latency but don't care about slightly older data, you can use this topology. In this case, you add the AS OF SYSTEM TIME clause in your reads, which then avoids the round trip to the leaseholder, and data is served locally. Writes would still need a multi-region consensus. You should not use this topology if you need strong consistency. Please refer to the *CAP theorem* section in *Chapter 1, CockroachDB – A Brief Introduction*, if you want to understand what consistency here means.

Let's create a sample table to understand how this works:

```
CREATE TABLE local_attractions (
        id UUID NOT NULL DEFAULT gen_random_uuid(),
        name STRING NOT NULL,
        address STRING NOT NULL,
        city STRING NOT NULL,
        PRIMARY KEY (id ASC)
);
```

Here, in the SELECT query, you should use AS OF SYSTEM TIME follower_read_timestamp(). The follower_read_timestamp() function returns the TIMESTAMP data type with statement_timestamp() – the 4.8s value:

```
SELECT city FROM local_attractions
    AS OF SYSTEM TIME follower_read_timestamp()
            WHERE city = 'san francisco';
```

Since the data is always retrieved locally, without involving the leaseholder, you might get stale or older data. Once again, do not use this topology if you need strong consistency.

Next, we will look at the follow-the-workload topology.

Follow-the-workload

This is the default topology if you don't use any of the previous ones. This topology works well if a given table is active in a single region, which means clients are doing reads and writes that are in the same region. Here, the read latency will be low in the active region and it will be higher in non-active regions, as the leaseholder will be in the active region. Writes still need a multi-region consensus and can be slower.

The following is a table that should help you to decide which topology might be relevant for your database workload:

Pattern	Latency	Resiliency	Configuration
Geo-partitioned replicas	Fast regional reads and writes	Can withstand one availability zone failure per partition	Geo-partitioned table Partition replicas pinned to regions
Geo-partitioned leaseholders	Fast regional reads Slower cross-region writes	Can withstand one regional failure	Geo-partitioned table Partition replicas spread across regions Partition leaseholders pinned to regions
Duplicate indexes	Fast regional reads Very slow cross-region writes, as even indexes must be replicated	Can withstand one regional failure	Multiple identical indexes Index replicas spread across regions Index leaseholders pinned to regions
Follower reads	Fast regional reads for stale data Slower cross-region writes	Can withstand one regional failure	Applications configured to use follower reads
Follow-the-workload	Fast regional reads, in the active region Slower cross-region reads, in non-active regions Slower cross-region writes	Can withstand one regional failure	No configuration required as this is the default

Figure 4.6 – Topology cheat sheet

In this section, we learned about various geo-partitioning topologies and how to configure them in CockroachDB. Based on latency, data consistency, and resiliency requirements, we should select the appropriate topology.

Summary

In this chapter, we learned what geo-partitioning is and why it is useful to geo-partition your data. Then, we covered all the important jargon in the cloud world, especially how each major cloud provider has defined regions and availability zones. We later discussed how to configure various multi-region topologies based on your application requirements.

In the next chapter, we will go over fault tolerance and auto-recovery with CockroachDB.

5
Fault Tolerance and Auto-Rebalancing

In *Chapter 4, Geo-Partitioning*, we learned about what **geo-partitioning** is, why we need it, and how it's supported in **CockroachDB**.

In this chapter, we will discuss what **fault tolerance** and **auto-rebalancing** are and how CockroachDB provides these features. We will also learn about multi-node failure scenarios and how to recover from them.

Fault tolerance refers to how CockroachDB copes with various types of failures. Auto-rebalancing in general is the ability to adapt and increase or decrease the number of nodes in a cluster to avoid hotspots. We will discuss auto-rebalancing with specific examples that you can also try.

The following topics will be covered in this chapter:

- Achieving fault tolerance
- Automatic rebalancing
- Recovering from multi-node failures

Technical requirements

We are going to discuss fault tolerance using an experiment in this chapter that will require you to have CockroachDB installed. If you still haven't done so, please refer to the *Technical requirements* section in *Chapter 2, How Does CockroachDB Work Internally?*.

Achieving fault tolerance

Fault tolerance is the ability to continue to operate even in the case of a system, network, or storage failure. This feature is critical to avoid data loss and for the continuity of your business. Whenever a node goes down or becomes *incommunicado*, the cluster automatically rebalances the number of replicas among remaining active nodes and continues to serve read and write traffic.

It is important to understand how many node failures you want to withstand, as based on that, you must decide how many nodes should be in your cluster. For example, in a cluster of three nodes, the cluster can withstand one node failure when the **replication factor** is three. In a cluster of seven nodes, the cluster can withstand two node failures when the replication factor is five.

Next, we will learn about having fault tolerance at the storage layer. After that, we will go over an example to understand fault tolerance using a six-node CockroachDB cluster. Finally, we will observe how CockroachDB rebalances data whenever a new node is added to the cluster.

Achieving fault tolerance at the storage layer

Although CockroachDB can help with fault tolerance and recovery at the database level, it is also very important to have the same at the storage layer as well.

Disk corruption, the spread of corruption to replicas, and data loss are some of the issues you might see at the storage layer. Fortunately, many cloud providers already provide storage options to avoid this.

For example, **Amazon Elastic Block Store** (**EBS**) provides persistent block storage volumes that can be used with **Amazon Elastic Compute Cloud** (**EC2**) instances. EBS volumes are automatically replicated within an availability zone, which provides high durability and availability.

Similarly, you can use persistent disks on **Google Cloud Platform** (**GCP**). Data on each persistent disk is distributed onto several physical disks. **Compute Engine** is responsible for managing these physical disks. Compute Engine is also responsible for data distribution, data redundancy, and increased performance. Make sure you pick the right option for storage, one that provides high availability and durability with optimal performance.

Now, let's look at an example offered by Cockroach Labs in order to understand how CockroachDB continues to operate in case of failures.

Working example of fault tolerance at play

All the commands used in this chapter will also be shared in the following GitHub repository: `https://github.com/PacktPublishing/Getting-Started-with-CockroachDB`:

1. **Setting up a six-node cluster**: The following is the command for starting the first node:

    ```
    Start node 1
    $ cockroach start \
    --insecure \
    --store=fault-node1 \
    --listen-addr=localhost:26257 \
    --http-addr=localhost:8080 \
    --join=localhost:26257,localhost:26258,localhost:26259 \
    --background
    ```

    Start another five nodes in the same cluster with unique listening addresses and ports.

    For example, you can use the following combinations of listening addresses and HTTP ports for the rest of the nodes: `26258/8081, 26259/8082, 26260/8083, 26261/8084, 26262/8085`. Please refer to `https://github.com/PacktPublishing/Getting-Started-with-CockroachDB` if you want complete commands to start the rest of the five nodes.

2. **Initializing the cluster**: You can initialize the cluster with the following command:

```
$ cockroach init \
--insecure \
--host=localhost:26257
Output
Cluster successfully initialized
```

3. **Setting up the load balancer**: The load balancer helps with spreading the requests between all the nodes. Cockroach Labs recommend using HAProxy as the load balancer. The following are the instructions for installing HAProxy on the different operating systems:

```
Mac
$ brew install haproxy
Ubuntu
$ sudo apt-get update
$ sudo apt-get install haproxy
Linux
$ yum install haproxy
```

The following command generates an `haproxy.cfg` file that you can use to provide load balancing for your CockroachDB cluster:

```
$cockroach gen haproxy \
--insecure \
--host=localhost \
--port=26257
```

You can view the contents of the generated file using the following command:

```
$ cat haproxy.cfg
```

In `haproxy.cfg`, change `bind :26257` to `bind :26000`. This port is used by HAProxy to accept requests and it should not coincide with any of the previously used ports.

Before you change the port to 26000, please make sure that no other process is already using it with the help of the following command:

```
$ lsof -i -P -n | grep 26000
```

The preceding command will return empty output if there are no other processes using port 26000. If 26000 is already occupied, pick some other port for HAProxy:

```
$ sed -i.saved 's/^    bind :26257/    bind :26000/'
haproxy.cfg
```

4. Next, start HAProxy by providing the haproxy.cfg file in the input as follows:

```
$ haproxy -f haproxy.cfg &
```

Now you can check port 26000 to make sure it's used by HAProxy:

```
$ lsof -i -P -n | grep 26000
haproxy   26517 kishen     5u  IPv4 0x4067c1a3fa14b465
0t0  TCP *:26000 (LISTEN)
```

5. **Running a sample workload**: You can use the cockroach workload command to run CockroachDB's built-in version of the **Yahoo! Cloud Servicing Benchmark (YCSB)**, which simulates multiple client connections that perform a combination of read and write operations.

6. Load the ycsb schema and data through HAProxy as follows:

```
$ cockroach workload init ycsb --splits=50 \
'postgresql://root@localhost:26000?sslmode=disable'
Output

I210516 10:00:04.066268 1 workload/workloadsql/dataload.
go:140  imported usertable (5s, 10000 rows)
I210516 10:00:04.084302 1 workload/workloadsql/
workloadsql.go:113  starting 50 splits
```

This workload creates a new ycsb database and a usertable table in that database and inserts data into the table. The --splits flag tells the workload to manually split ranges at least 50 times.

Now, let's run the `ycsb` workload as follows:

```
$ cockroach workload run ycsb \
--duration=30m \
--concurrency=5 \
--max-rate=500 \
--tolerate-errors \
'postgresql://root@localhost:26000?sslmode=disable'
```

The preceding command initiates 5 concurrent client workloads for 30 minutes but limits the total load to 500 operations per second.

Per-operation statistics are printed to standard output every second.

After the specified duration, the workload will stop, and you will see the summary printed in the standard output, as shown here:

```
 _elapsed___errors_____ops(total)___ops/sec(cum)__
avg(ms)__p50(ms)__p95(ms)__p99(ms)_pMax(ms)__total
   1200.0s          0          567529          472.9       1.5
1.0        2.9        5.5      436.2   read
```

```
 _elapsed___errors_____ops(total)___ops/sec(cum)__
avg(ms)__p50(ms)__p95(ms)__p99(ms)_pMax(ms)__total
   1200.0s          0           29755           24.8      81.0
75.5      151.0      192.9    486.5   update
```

```
 _elapsed___errors_____ops(total)___ops/sec(cum)__
avg(ms)__p50(ms)__p95(ms)__p99(ms)_pMax(ms)__result
   1200.0s          0          597284          497.7       5.4
1.0       41.9      109.1    486.5
```

Open the CockroachDB web UI at `http://localhost:8080`. This web UI can also be opened at ports `8081`, `8082`, `8083`, `8084`, and `8085`.

To check the SQL queries being executed, click on **Metrics** on the left, and hover over the **SQL Queries** graph at the top, as shown in the following screenshot:

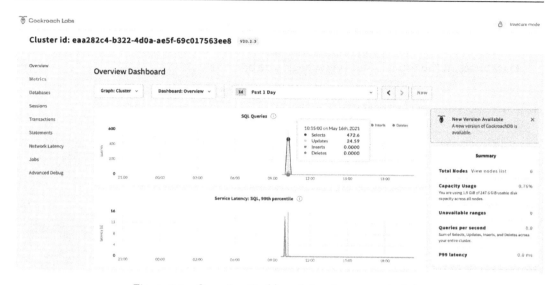

Figure 5.1 – Overview Dashboard showing query statistics

In the CockroachDB web UI, to check the client connections from the load generator, select **SQL Dashboard** and hover the cursor over the **SQL Connections** graph, as shown in the following screenshot:

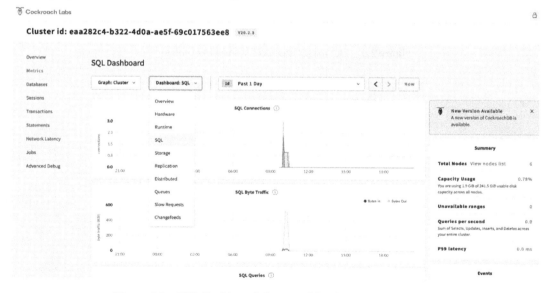

Figure 5.2 – SQL Dashboard showing SQL Connections metrics

You will notice three client connections from the load generator.

To see more details about the `ycsb` database and the `usertable` table, click **Databases** in the top left and then check `ycsb`:

Figure 5.3 – Databases dashboard showing tables

You can also view the schema of `usertable` by clicking the table name as follows:

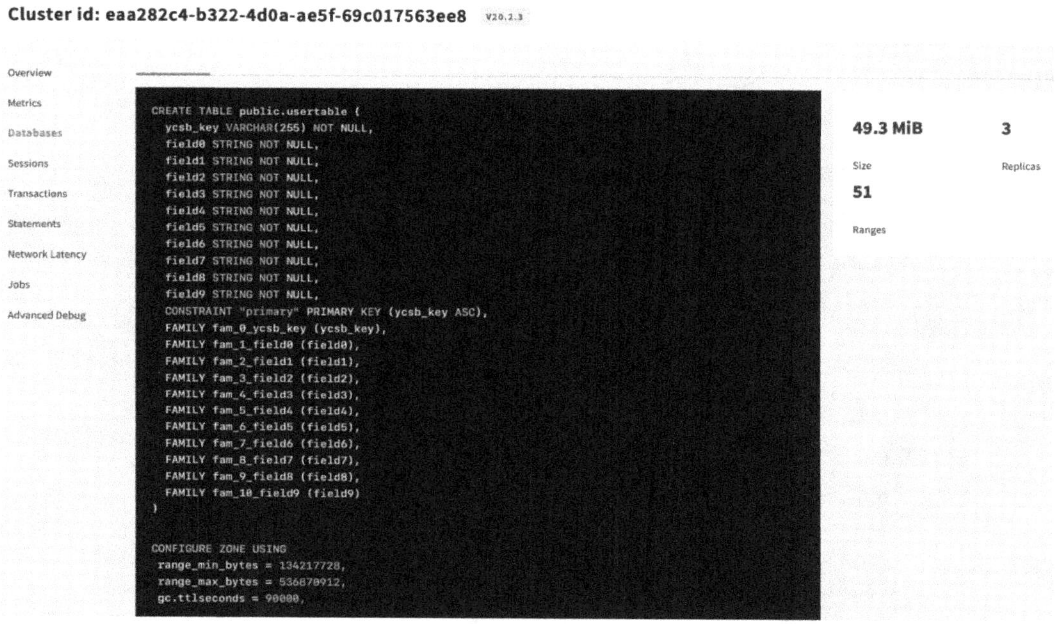

Cluster id: eaa282c4-b322-4d0a-ae5f-69c017563ee8 v20.2.3

Figure 5.4 – Viewing the table schemas

By default, CockroachDB replicates all data three times and balances it across all nodes. To see this balance, go to **Overview** and check the replica count across all nodes as follows:

Figure 5.5 – Overview dashboard showing Node Status and Replication Status

Next, we will look at simulating a single-node failure.

7. **Simulating a single-node failure**: By default, a node is considered dead when it doesn't respond for at least 5 minutes. After that, CockroachDB starts replicating any replicas that were on the dead node to other active nodes. In this setup, since we no longer want to wait for 5 minutes, we can change the time to 75 seconds, as shown in the following command:

```
$ cockroach sql \
--insecure \
--host=localhost:26000 \
--execute="SET CLUSTER SETTING server.time_until_store_
dead = '1m15s';"
Output
SET CLUSTER SETTING
Time: 278ms
```

To permanently bring a node down, we can use the `quit` command:

```
$ cockroach quit \
--insecure \
--host=localhost:26261
Output
Command "quit" is deprecated, see 'cockroach node drain'
instead to drain a
server without terminating the server process (which can
in turn be done using
an orchestration layer or a process manager, or by
sending a termination signal
directly).
node is draining... remaining: 38
node is draining... remaining: 0 (complete)
ok
```

8. **Keep checking the cluster health**: Go back to the CockroachDB console, click on **Metrics**, and check that the cluster continues to service data, despite one of the nodes being unavailable. Also, keep an eye on **Unavailable ranges 0** on the right-hand side panel, which should be zero throughout this setup.

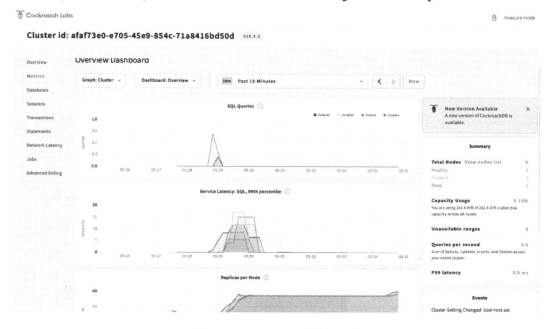

Figure 5.6 – Metrics dashboard

The current setup shows that when all the ranges are replicated three times, a three-node cluster can withstand a single-node failure without affecting overall availability.

9. **Watch the cluster self-heal**: Here you can see that data has been replicated from the unhealthy node to all the healthy nodes. So, now there are **0** under-replicated ranges and **0** unavailable ranges. This means the cluster is now ready to accept requests for any range, although one of the nodes is dead, as can be seen in the following screenshot:

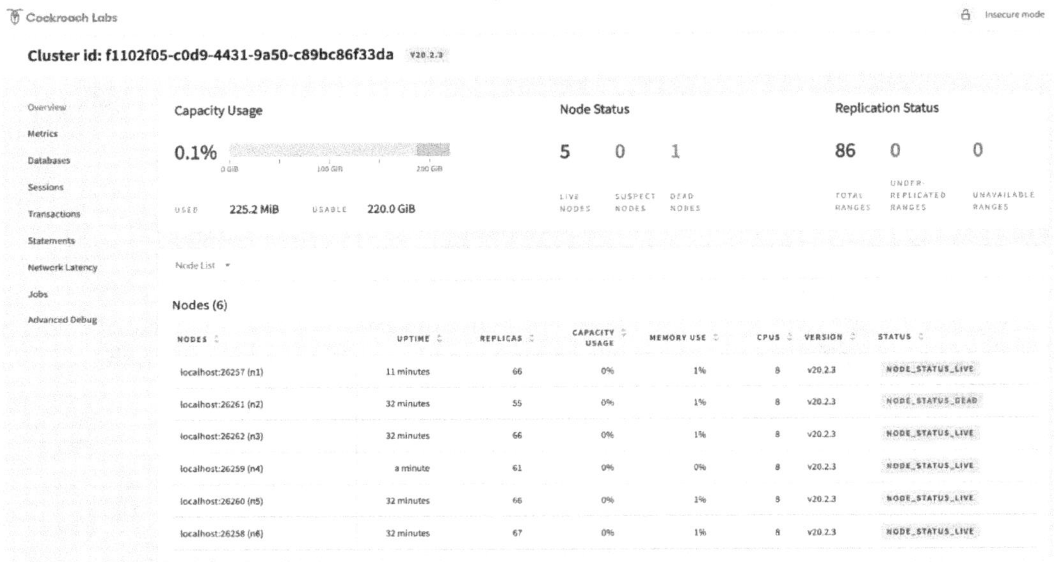

Figure 5.7 – Overview dashboard showing five live nodes and one dead node

As discussed in *step 7*, once a given node is considered *dead*, replicas from the dead node are moved to other active nodes. So, you should see an increase in the replication count for the active nodes. A CockroachDB cluster heals itself by re-replicating the replicas from the dead node to active nodes.

In the next section, we will learn about how CockroachDB automatically rebalances the replicas when we add more nodes.

Automatic rebalancing

Automatic rebalancing is the process of rebalancing the replicas from currently active nodes to newly added ones.

Now, let's take the cluster from the previous section, which had five active nodes and one dead node. In this section, we are going to observe how data is rebalanced once we add a new node to a cluster. The following screenshot shows the node replication status:

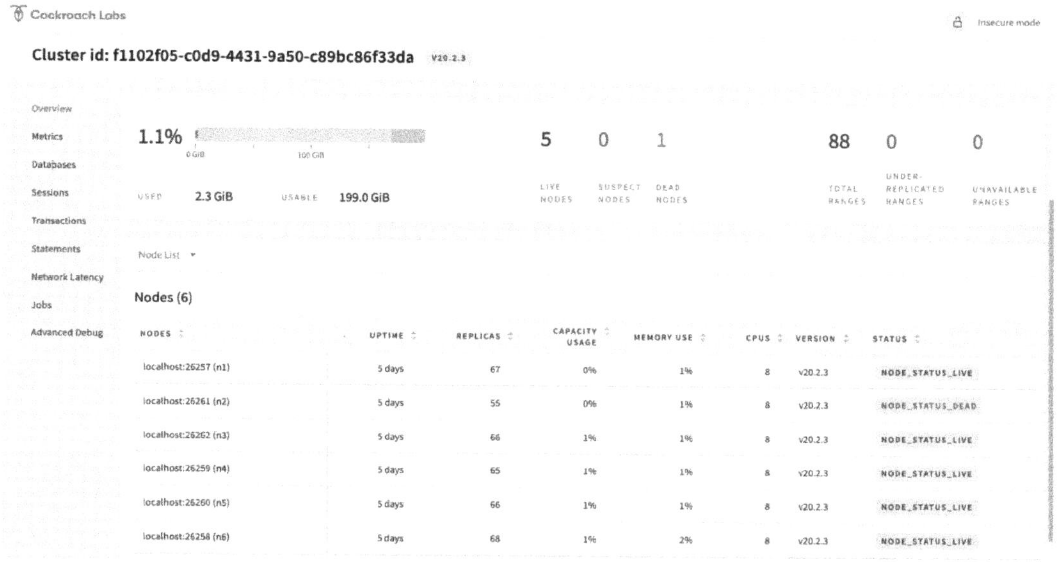

Figure 5.8 – Overview dashboard showing Node Status and Replication Status

In this cluster of five active nodes, let's add two more nodes as follows:

```
$ cockroach start \
--insecure \
--store=fault-node5 \
--listen-addr=localhost:26261 \
--http-addr=localhost:8084 \
--join=localhost:26257,localhost:26258,localhost:26259 \
--background

cockroach start \
--insecure \
--store=fault-node7 \
--listen-addr=localhost:26263 \
--http-addr=localhost:8086 \
--join=localhost:26257,localhost:26258,localhost:26259 \
--background
```

After some time, you should see that now there are **7** active nodes. The replication count should also come down for all the previous nodes, since the replicas are rebalanced between all the nodes in the cluster. The following figure shows the CockroachDB cluster with **7** active nodes:

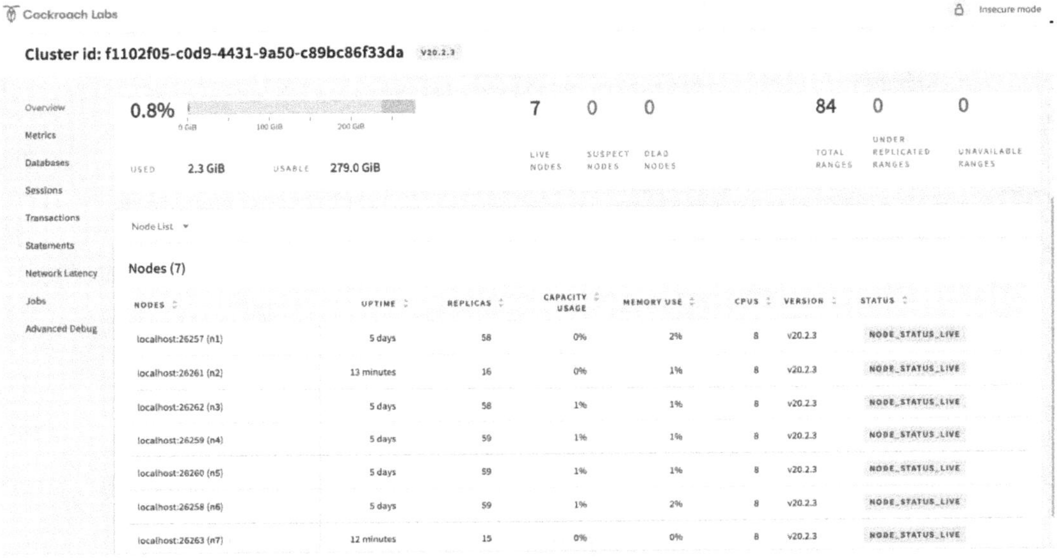

Figure 5.9 – Overview dashboard showing seven live nodes

In the next section, we will learn about multi-node failures.

Recovering from multi-node failures

If you want your cluster to withstand multi-node failures while continuously serving all the ranges, then you should ensure that you have enough active nodes available for all the replicas.

For example, in the previous section, we created a seven-node cluster and the replication count was three. If two nodes go down simultaneously, then some ranges will become unavailable, as there will not be a majority consensus if a given range is replicated in the two nodes that went down. So, if you want this seven-node cluster to withstand two node failures, you must increase the replication factor to five, so that there will still be a majority of 3/5 for some ranges that had replicas in the two nodes that went down. In general, a cluster can continue to serve all the ranges when *(replication factor – 1) / 2* nodes go down.

You can use the following command to change the replication factor to 5:

```
$ cockroach sql --execute="ALTER RANGE default CONFIGURE ZONE
USING num_replicas=5;" --insecure --host=localhost:26000
```

After the replication factor of five is consumed by the cluster and five replicas are created for each range, you can use cockroach quit to bring down any two nodes. After a few minutes, you can check that all the ranges are still available.

When you are trying the various configurations provided in this chapter, if things don't go as discussed, please refer to *Chapter 10, Troubleshooting Issues*, in order to understand the failure and how to fix it.

> **Note**
>
> Please refer to the cleanup.sh script at https://github.com/ PacktPublishing/Getting-Started-with-CockroachDB in order to clean up your cluster. You can also use **SIGKILL** to kill the processes of CockroachDB instances and HAProxy, and later you can manually delete the data folders.

Summary

In this chapter, we learned about fault tolerance, auto rebalancing, and how to recover from multi-node failures. We also went through a few configurations to understand how CockroachDB provides these features. Basically, fault tolerance gives us enough time for the **Site Reliability Engineering** (**SRE**) and DevOps teams to deal with node failures without any service disruption.

In the next chapter, we will learn about indexes and how they are implemented in CockroachDB.

6

How Indexes Work in CockroachDB

In *Chapter 5, Fault Tolerance and Auto-Rebalancing*, we learned about fault-tolerance and auto-recovery strategies in CockroachDB. In this chapter, we will learn everything about **indexes**, what they are, and how they improve query times.

Although indexes help improve the read performance, a wrong index can slow down the queries, including the writes, and take up more storage space. So, it is important to identify the query pattern and create appropriate indexes.

The following topics will be covered in this chapter:

- Introduction to indexes
- Different types of indexes
- Best practices while using indexes

Technical requirements

We will try creating various types of indexes in this chapter which would require you to have CockroachDB installed. If you still haven't done so, please refer to the *Technical requirements section in Chapter 2, How Does CockroachDB Work Internally?*.

Introduction to indexes

An index or a **database index** helps with returning the query results quickly, by avoiding full table scans. An index can be created for a specific table and can include one or more keys. **Keys** refer to the columns in the table. However, there will be extra space used to keep a separate sorted copy of indexed columns.

Let's take a simple example and see how an index works.

Consider a population table with the following columns and some sample values:

id	country	continent	population_in_millions
1	India	Asia	1378
2	USA	North America	331
3	South Africa	Africa	60
4	China	Asia	1400
5	Switzerland	Europe	8.5

Figure 6.1 – Population table

Now, let's say you just want to retrieve the list of populations for specific continents, for example:

```
SELECT population_in_millions, country FROM population WHERE
continent = "Asia";
```

Here, in order to find rows 1 and 4, which are countries in Asia, you would have to iterate through each of the rows in the table, which is called a **full table scan**.

Now, if you want to avoid a full table scan, you can create an index on the continent as follows:

```
CREATE INDEX ON population (continent);
```

Internally, CockroachDB keeps track of all the continents and keeps a mapping from a given continent to all its relevant rows, as shown here:

```
Africa -> (3)

Asia    -> ( 1, 4 )

Europe -> ( 5 )

North America -> (2 )
```

Now, if you run the same query, `SELECT population_in_millions, country FROM population WHERE continent = "Asia"`, once again, CockroachDB will identify that the filtering condition has the column `continent` and there is an index already available for that. So, in this case, based on the value `Asia`, it will directly retrieve rows (1 , 4) from the continent index and get relevant column values and return them. In this case, it has avoided a full table scan. Although this example only has five rows, the same concept is applicable even when a table contains millions of rows. So, in such cases, avoiding a full table scan can significantly improve the query performance. At the same time, writes tend to get a little slower as, with each write, even the index must be updated.

When you create an index on a column or a set of columns, CockroachDB internally makes a copy of the values of that column or columns and sorts them. So, whenever you execute a query that involves filters on indexed columns, a subset of rows is selected from the index first, rather than scanning the entire table. This improves the overall query performance.

Next, we are going to discuss various types of indexes that are available in CockroachDB.

Different types of indexes

Based on the query pattern and columns in the table, you should decide what kind of index is going to help with the performance.

The following are the types of indexes available in CockroachDB:

- Primary index
- Secondary index
- Hash-sharded index
- Duplicate indexes
- Inverted indexes
- Partial indexes
- Spatial indexes
- Table joins and indexes
- Best practices while using indexes

In the next set of subsections, we are going to discuss each type of index and when to use them, starting with the primary index.

Primary indexes

A **primary key** uniquely identifies a given row in a table. This means that the primary key is unique for a given row and duplicate values or NULLs are not allowed. An index created for a primary key is called a **primary index**.

Whenever you create a table in CockroachDB, it's recommended to have an explicit primary key, so that CockroachDB automatically creates an index for it, which can be used to filter the rows for better performance. Even if you don't create a primary key during table creation, CockroachDB by default creates a primary key called rowid, which will have a unique value for each row, but its performance will not be as good as that of the primary key.

Let's understand how indexes work with an example, where we are going to create a database and a table with a primary key:

1. Create a database called test:

    ```
    root@localhost:26257/defaultdb> CREATE DATABASE IF NOT
    EXISTS test;
    CREATE DATABASE


    Time: 279ms total (execution 279ms / network 0ms)
    Switch to the database 'test'.
    root@localhost:26257/defaultdb> use test;
    SET


    Time: 116ms total (execution 115ms / network 0ms)
    ```

2. Create a table called accounts with id being the primary key:

    ```
    root@localhost:26257/test> CREATE TABLE accounts (
            id UUID PRIMARY KEY,
            name string,
            balance INT8
    );
    CREATE TABLE
    Time: 195ms total (execution 195ms / network 0ms)
    ```

3. Now, we will look at the indexes created for the `accounts` table using the `SHOW INDEXES` command:

```
SHOW INDEXES FROM accounts;
    table_name | index_name | non_unique | seq_in_index |
column_name | direction | storing | implicit
-------------+------------+------------+--------------+--
-----------+------------+---------+-----------
    accounts    |  primary   |    false   |              |
              1 | id         |            |
ASC             |    false   |   false    |
(1 row)


Time: 7ms total (execution 7ms / network 0ms)
```

In the next section, we will learn about hash-sharded indexes that are used in relation to sequences.

1. In order to understand how this primary key index helps with query performance, we can use `EXPLAIN` to look at the statement plans.

2. If you are retrieving all the accounts without any filters, obviously a full scan is required as we must return all the rows:

```
root@localhost:26257/test> explain select * from
accounts;
    tree    |        field        |     description
------------+---------------------+--------------------
            | distribution        | full
            | vectorized          | false
    scan    |                     |
            | estimated row count | 1
            | table               | accounts@primary
            | spans               | FULL SCAN
(6 rows)
```

3. Now if you want to retrieve just one row based on the ID, you can avoid a full table scan, since CockroachDB has already indexed the `id` column.

4. As you can see in the following `explain` statement, within the spans, now we no longer do a full table scan:

```
root@localhost:26257/test> explain select * FROM accounts
where id = '123e4567-e89b-12d3-a456-426655440000';
     tree     |           field            |      description
--------------+----------------------------+--------------------
              | distribution               | local
              | vectorized                 | false
     scan     |                            |
              | estimated row count        | 1
              | table                      | accounts@primary
              | spans                      | [/'123e4567-
e89b-12d3-a456-426655440000' - /'123e4567-e89b-
12d3-a456-426655440000']
(6 rows)
```

If multiple columns are used in queries, you should also consider creating a composite primary key that includes all the columns that are often used together.

In the next section, we will learn about secondary indexes.

Secondary indexes

A **secondary index** is an index that you create on non-primary columns. If your query involves retrieving a column that's not a primary key and you want to improve the query performance, you can create secondary indexes. Any index that you create on a non-primary key is called a secondary index, and duplicate values are allowed for secondary indexes. For the `test.accounts` table, if the query contains a non-primary column such as `name`, then we would still need a full table scan. Let's try this with an example, where we will just use a non-primary column in the query:

```
root@localhost:26257/test> explain select name FROM accounts
where name = 'crdb' ;
     tree     |           field            |      description
--------------+----------------------------+--------------------
              | distribution               | full
              | vectorized                 | false
     filter   |                            |
              | filter                     | name = 'crdb'
```

```
      └── scan  |                              |
                | estimated row count | 1
                | table               | accounts@primary
                | spans               | FULL SCAN
(8 rows)
```

Since we are now filtering on a non-primary column, CockroachDB must inspect each row and apply a filtering condition, and the index on the primary key id doesn't help here. So, let's create one more index on the column name:

```
root@localhost:26257/test> CREATE INDEX on accounts ( name );
CREATE INDEX


Time: 2.053s total (execution 0.256s / network 1.797s)
```

Whenever you create a secondary index, CockroachDB automatically creates a composite index including the primary key. Also, the index on the column name is called a secondary index:

```
root@localhost:26257/test> show indexes from accounts;
    table_name |       index_name       | non_unique |
seq_in_index | column_name | direction | storing | implicit
-------------+---------------------+-----------+---------------+-
-----------+----------+---------+-----------
    accounts  | primary
|      false  |                              1
| id                        |
ASC                | false   | false
    accounts  | accounts_name_idx
|      true   |                              1
| name                      |
ASC                | false   | false
    accounts  | accounts_name_idx
|      true   |                              2
| id                        |
ASC                | false   | true
(3 rows)
```

Now if you run the previous query, you should see that the full table scan is avoided because of the new index that we have created:

```
root@localhost:26257/test> explain select name from accounts
where name = 'crdb' ;
    tree     |        field       |       description
-------------+--------------------+--------------------------
             | distribution       | local
             | vectorized         | false
      Scan   |                    |
             | estimated row count| 1
             | table              | accounts@accounts_name_idx
             | spans              | [/'crdb' - /'crdb']
  (6 rows)


Time: 1ms total (execution 1ms / network 0ms)
```

Hash-sharded indexes

Hash-sharded indexes can improve query performance when you must create an index on a column that's a sequence. Hash-sharded indexes evenly spread the traffic to a sequential range across multiple ranges to avoid hotspots for any given range. Since this is a new experimental feature, the implementation and overall performance might change over time. Let's begin:

1. Within the client session, you have to first enable this feature as shown in the following code block:

```
root@localhost:26257/test> set experimental_enable_hash_
sharded_indexes = ON;
SET


Time: 1ms total (execution 0ms / network 0ms)
```

2. Let's create a table called `customers` with integer and string data types. Here, the `id` column is supposed to be a sequence:

```
root@localhost:26257/test> create TABLE customers ( id
int PRIMARY KEY, name string);
CREATE TABLE


Time: 160ms total (execution 160ms / network 0ms)
```

3. Now, let's create the hash-sharded index for this primary key, as shown in the following code:

```
root@localhost:26257/test> ALTER TABLE customers ALTER
PRIMARY KEY USING COLUMNS (id) USING HASH WITH BUCKET_
COUNT = 10;
NOTICE: primary key changes are finalized asynchronously;
further schema changes on this table may be restricted
until the job completes
ALTER TABLE


Time: 4.551s total (execution 0.297s / network 4.253s)
```

When you create a hash-sharded index, CockroachDB creates n_ buckets computed columns, shards the primary index ID into n_ buckets number of shards, and then stores each index shard in the underlying key-value store with one of the computed column's hashes as its prefix.

4. Let's look at how the indexes on the `customers` table look now:

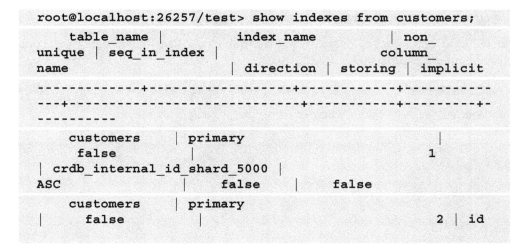

```
root@localhost:26257/test> show indexes from customers;
    table_name |        index_name         | non_
unique | seq_in_index |             column_
name                | direction | storing | implicit
-------------+-------------------+-----------+----------
---+-------------------------------+-----------+---------+-
----------
    customers  | primary                           |
      false    |                              1
 | crdb_internal_id_shard_5000 |
ASC              |     false   |     false
    customers  | primary
 |    false    |                           2 | id
```

		ASC		false		false
customers		customers_id_key				
	false				1	
id						
		ASC		false		false
customers		customers_id_key				
	false				2	
	crdb_internal_id_shard_5000					
ASC			false		true	

```
(4 rows)
```

```
Time: 34ms total (execution 26ms / network 8ms)
```

You can create a hash-sharded secondary index as well.

Duplicate indexes

Duplicate indexes improve the read performance. Please refer to *Chapter 4, Geo-Partitioning*, where we discussed duplicate indexes and how they work internally.

Next, we will learn about inverted indexes.

Inverted indexes

Inverted indexes store the mapping of values within JSONB, arrays, and spatial data to the row that holds that value. For example, if you have a column where you are storing a JSON document, and let's say that JSON document contains a key called country, then you can add a WHERE clause in your query, where you can say get me all the rows that have country:USA and country:Canada.

Inverted indexes filter on components of tokenizable data. The JSONB data type is built on two structures that can be tokenized:

- **Objects** – Collections of key and value pairs where each key-value pair is a token

- **Arrays** – Lists of values where every value in the array is a token

Let's look at the following JSON document:

```
"student": [
    {
        "id":"01",
        "firstname": "Steve",
        "lastname": "Jobs"
    },
    {
        "id":"02",
        "firstname": "Steve",
        "lastname": "Wozniak"
    }
]
```

Now, the inverted index for the preceding JSON will have an entry for each component, which maps to the original document as follows:

```
"student"  :     "id" : "01"
"student"  :     "firstname" : "Steve"
"student"  :     "lastname" : "Jobs"
"student"  :     "id" : "02"
"student"  :     "lastname": "Wozniak"
```

Now you can search the JSON document based on student ID, student first name, student last name, and so on.

Partial indexes

A **partial index** is typically created based on a Boolean expression. CockroachDB internally indexes the columns and rows that evaluate to true for a given expression.

Let's understand partial indexes with an example:

1. First, we will create the table books with a few columns:

```
root@localhost:26257/test> create table books ( id int,
title string, author string, price float );
CREATE TABLE


Time: 270ms total (execution 269ms / network 1ms)
```

2. Let's create the partial index based on the price of the book. Here, we are creating an index for all the books that are priced more than $50:

```
root@localhost:26257/test> CREATE INDEX ON books (id,
title, author) WHERE price > 50.00;
CREATE INDEX


Time: 2.596s total (execution 0.314s / network 2.282s)
```

3. Now, whenever you use a filtering condition that matches with the one in the partial index, a partial index will be used to retrieve a subset of rows:

```
root@localhost:26257/test> explain select id, name,
author from books where price > 50.0;
        tree |                    field
             |
             description
-------+--------------------+----------------------------
---------------------
             | distribution                     | full
             | vectorized                       |
false
    scan |
         |
             | estimated row count | 1
             | table
         | books@books_id_name_author_idx (partial
index)
             | spans
         | FULL SCAN
  (6 rows)


Time: 2ms total (execution 1ms / network 1ms)
```

Partial indexes improve the query performance in the following ways:

- They contain fewer rows than full indexes. During read queries, only rows in the partial index are scanned, if there is a match in the filtering condition. Since partial indexes contain fewer rows compared to regular indexes, we will be scanning fewer rows, so it performs better than a regular index.

- Write queries on tables with a partial index only perform an index write when the rows inserted satisfy the partial index predicate, unlike regular indexes, which are updated during every write.

In the next section, we will learn about spatial indexes.

Spatial indexes

Spatial indexes were introduced in the v20.2.16 version, are used to store information about spatial objects, and mostly work with two-dimensional data types such as GEOMETRY and GEOGRAPHY. A spatial object holds information about a geographical location in the form of an object. Here, an object can be a point, a line, a polygon, or an area.

A spatial index is a special type of inverted index. A spatial index maps from a cell in a quadtree to one or more shapes whose coverings include that cell. Each cell can be part of multiple shapes, where a given cell represents a location.

Spatial indexes are useful in the following situations:

- We are filtering based on spatial predicate functions, for example, ST_COVERS(*), ST_CONTAINS, ST_Equals, ST_Overlaps, and so on.

- Joins that involve columns that store spatial objects.

CockroachDB uses the S2 geometry library (https://s2geometry.io/) to divide the space being indexed and stores the information in a **quadtree** data structure.

A quadtree is a tree data structure in which each internal node has exactly four children. Each cell in a quadtree has information about four child cells in the next level. In the following diagram, you can see an example of how an image can be represented using a quadtree data structure:

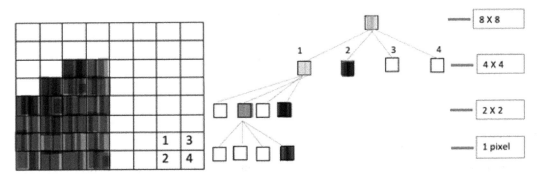

Figure 6.2 - Image representation using a Quadtree data structure

Following are some examples of creating spatial indexes.

First, let's create a table with GEOGRAPHY and GEOMETRY columns:

```
root@localhost:26258/test> CREATE TABLE geo_table (
    id UUID PRIMARY KEY,
    geog GEOGRAPHY(GEOMETRY,4326) NULL,
    geom GEOMETRY(GEOMETRY,3857) NULL
);
CREATE TABLE


Time: 151ms total (execution 149ms / network 2ms)
```

Following is an example of creating a spatial index on a GEOMETRY object with default settings:

```
root@localhost:26258/test> CREATE INDEX geom_idx_1 ON geo_table
USING GIST(geom);
CREATE INDEX


Time: 1.647s total (execution 0.137s / network 1.511s)
```

Following is an example of creating a spatial index on a GEOGRAPHY object with default settings:

```
root@localhost:26258/test> CREATE INDEX geog_idx_1 ON geo_table
USING GIST(geog);
CREATE INDEX


Time: 1.709s total (execution 0.144s / network 1.564s)
```

Fine-tuning spatial indexes is beyond the scope of this book and will be covered in subsequent editions.

Next, we will learn how to improve the performance of queries that involve table joins.

Table joins and indexes

Indexes are useful even when you are joining tables. You can inspect the fields that are used in filtering conditions and create appropriate indexes to avoid a full scan of other indexes.

For example, let's look at two tables: customers and purchase_orders.

The customers table stores the information about the customers as follows:

```
root@localhost:26258/test> CREATE TABLE customers (
        id UUID PRIMARY KEY,
        name STRING,
        email STRING,
        phone STRING
);
CREATE TABLE


Time: 209ms total (execution 209ms / network 0ms)
```

The purchase_orders table stores information about the purchase orders made by the customers. Here, the customer_id column references the id column of the customers table:

```
root@localhost:26258/test> CREATE TABLE purchase_orders (


        id UUID PRIMARY KEY,
        customer_id UUID NOT NULL REFERENCES customers ( id
```

```
),
        n_of_items INT,
        total_price DECIMAL(10,2)

);
CREATE TABLE


Time: 843ms total (execution 264ms / network 579ms)
```

Now, let's say you want to know the name of all the customers who have purchased more than five items. Following is the explain plan for this:

```
root@localhost:26258/test> explain select n_of_items, name from
purchase_orders    INNER JOIN customers ON purchase_orders.
customer_id = customers.id and n_of_items > 5;
              tree              |             field
                 |        description
-----------------+----------------------+----------------------
----
                                         |
distribution                   | full
                                         |
vectorized                      | false
     hash join                  |
                                         |
     |                          |
equality                                 | (customer_id) = (id)
     |                          | right cols are key       |
     ├── filter                 |
                                         |
     |          |              |
filter                                   | n_of_items > 5
     |          └── scan        |
  |                                                         |
     |                          | estimated row count | 1
     |                          |
table                                    | purchase_orders@
primary
     |                          |
spans                                    | FULL SCAN
```

```
└── scan              |
                                    |
                         | estimated row count |
1
                         |
table                    | customers@primary
                         |
spans                    | FULL SCAN
```

As you can see, we are making use of primary indexes on both the customers and purchase_orders tables. But we also have a filtering condition for which we are using the column n_of_items. We can further improve the query performance by adding one more index for the n_of_items column:

```
CREATE INDEX ON purchase_orders (n_of_items) STORING (customer_
id);
```

Now, let's once again look at the explain plan for the previous query:

```
root@localhost:26258/test> explain select n_of_items, name from
purchase_orders    INNER JOIN customers ON purchase_orders.
customer_id = customers.id and n_of_items > 5;
        tree          |              field
         |                                  description
------------+----------------------+--------------------
-----------------------------
                      | distribution                          |
full
                      |
vectorized            | false
    hash join
 |                                               |
    |             |
equality                         | (customer_id) = (id)
    |             | right cols are key     |
 ├── scan
 |                                          |
    |             | estimated row count | 1
    |             |
table                            | purchase_orders@
purchase_orders_n_of_items_idx
```

```
spans                                    | [/6 - ]
     └── scan

                             | estimated row count | 1

table                                    | customers@primary

spans                                    | FULL SCAN
```

As you can see, now we no longer do a full scan of the `purchase_orders@primary` index. So, based on the columns used in filtering conditions during table joins, you can create appropriate indexes.

Next, we will go over some of the best practices to consider related to indexes that can improve the query performance.

Best practices while using indexes

Whenever you are using indexes, you can follow certain guidelines to make sure you get the best performance for your queries. Following are some of the key points to remember:

- Avoid creating an index on a sequence. Due to the nature of how the columns are sharded, sometimes we can have range hotspots, where most of the requests are coming for the same range, which can slow down the query. It would be best to use UUIDs or randomly generated keys. If you must create an index on a column that is sequential in nature, you should use hash-sharded indexes, as discussed in the *Hash-sharded indexes* section.

- If you are using multiple columns in your WHERE clause or in your ORDER BY clause, you should consider creating an index for all these columns.

- In your WHERE clause, make sure to have filters that are more restrictive before the ones that are a bit more generic. For example, = and IN should come before LIKE, >, ! =, and so on.

- You should drop the indexes that are not getting used. This will improve the write performance, as fewer indexes will have to be updated. Right now, there is no easy way to know the unused indexes. This requires manually going through the logical plans and identifying the indexes that are not getting used.

You can use DROP INDEX to drop a specific index.

For example, let's drop an index created previously, in the *Partial indexes* section:

```
root@localhost:26258/test> DROP INDEX books@books_id_name_
author_idx;
NOTICE: the data for dropped indexes is reclaimed
asynchronously
HINT: The reclamation delay can be customized in the zone
configuration for the table.
DROP INDEX


Time: 1.659s total (execution 0.178s / network 1.482s)
```

After this, if you execute SHOW INDEXES, you should not see the books@books_id_
name_author_idx index:

```
root@localhost:26258/test> SHOW INDEXES FROM books;
    table_name | index_name | non_unique | seq_in_index |
column_name | direction | storing | implicit
-------------+------------+------------+--------------+--------
-----+-----------+---------+-----------
    books            | primary           |      false
    |                       1 | rowid                          |
ASC              |     false    |     false
(1 row)


Time: 17ms total (execution 17ms / network 0ms)
```

We have discussed various types of indexes other than primary and secondary. Make sure you understand these specialized indexes and use them appropriately. If a given query is using an index but is still slow, perhaps you should investigate further to see if that index makes sense for the query or if some other type of index would better improve the performance.

You can also select a specific index in the query if you think that's going to improve the performance.

Summary

In this chapter, we learned about indexes, several special types of indexes, how they work internally, and the best practices for maximum query performance. It is important to understand the columns in your tables and query patterns and pick relevant indexes for maximum query performance.

In the next chapter, we will learn about high availability and how to deploy CockroachDB in order to achieve zero downtime and to make it highly available.

Section 3: Working with CockroachDB

This section introduces you to the practical aspects of managing CockroachDB as a database service. Managing schemas, monitoring the CockroachDB cluster using the UI, securing CockroachDB workloads, troubleshooting issues, conducting performance benchmarks, and migrating from a traditional database to CockroachDB will all be covered in this section.

This section comprises the following chapters:

- *Chapter 7, Schema Creation and Management*
- *Chapter 8, Exploring the Admin User Interface*
- *Chapter 9, An Overview Of Security Aspects*
- *Chapter 10, Troubleshooting Issues*
- *Chapter 11, Performance Benchmarking and Migration*
- *Appendix: Bibliography and Additional Resources*

7
Schema Creation and Management

In *Chapter 6, How Indexes Work in CockroachDB*, we learned what indexes are, how they are useful in improving query performance, the various types of indexes that are supported in CockroachDB, and the best practices while using indexes.

In this chapter, we will go through the syntax for various **Structured Query Language (SQL)** operations. Though we have learned some of the syntaxes throughout other chapters, it's useful to have all of them in a single place. Throughout this chapter, only commonly used query options are included, and some of the experimental and enterprise-only features have been left out.

The following topics will be covered in this chapter:

- **DDL**
- **DML**
- **DQL**
- Supported data types
- Column-level constraints
- Table joins

- Using sequences
- Managing schema changes

Technical requirements

We will need at least a single-node CockroachDB cluster to try some of the queries discussed in this chapter. So, please refer to the *Technical requirements* section of *Chapter 2, How Does CockroachDB Work Internally?*

If you want to try this on a larger CockroachDB cluster, then please refer to the *Working example of fault tolerance at play* section of *Chapter 5, Fault Tolerance and Auto-Rebalancing*, where we create a six-node cluster.

Also, if you are not sure about some SQL operations and how to use them, you can just use the CockroachDB **Help** option to get more information.

For example, if you want to know about all the options available with CREATE TABLE, you can just try \h CREATE TABLE in the SQL client console, as shown in the following code snippet:

```
root@localhost:26258/test> \h CREATE TABLE
```

You can try \h with any of the commands to get more detailed information.

In the first section, we will learn about DDL statements.

DDL

DDL statements are mainly responsible for creating, altering, and dropping tables, indexes, and users. DDL statements typically comprise CREATE, ALTER, and DROP operations. In this section, we will go over various DDL operations and their syntax, starting with CREATE.

CREATE

CREATE is the keyword used for creating something new such as a database, schema, table, view, or user.

A CREATE DATABASE statement accepts the following parameters:

- IF NOT EXISTS: Creates a new database, only if the database with the same name does not exist previously
- database_name: Name of the database to be created

The following databases are created by default and are used internally by CockroachDB:

- `postgres`: Empty database that is provided for compatibility with Postgres clients
- `system`: A database that contains CockroachDB metadata and that is read-only

The following default database is used for default connections:

- `defaultdb`: Used when a client doesn't specify a database in connection parameters

The following default databases are used for demonstration purposes:

- `movr`: Sample database with users, vehicles, and rides for vehicle-sharing apps
- `startrek`: A database that contains quotes from *Star Trek* episodes

Some of these databases are used internally by CockroachDB, and some are there with the sample schema maintained by *Cockroach Labs*, so it is advisable not to use any of these preloaded databases and to instead create separate ones for your application.

Let's look at a specific example of creating a database. You can view the code here:

```
CREATE DATABASE users PRIMARY REGION "us-central1" REGIONS
"us-east1", "us-central1", "us-west1" SURVIVE REGION FAILURE;
```

The preceding statement will create a database with the primary region being `"us-central1"` and with `"us-east1"`, `"us-central1"`, and `"us-west1"` database regions.

Next, we will look at the syntax of the `CREATE TABLE` statement.

CREATE TABLE syntax

The `CREATE TABLE` statement accepts the following parameters:

- `IF NOT EXISTS`: Creates a table if a table with the same name doesn't exist already.
- `table_name`: Name of the table.
- `column_def`: Column definition that includes a column name and a data type. Please refer to the *Column-level constraints* section to explore all the constraints you can provide for a column.
- `index_def`: Comma-separated list of index definitions.

- `family_def`: List of column family definitions. A comma is used as a separator. A `Column` family is stored as a single key-value pair.

- `table_constraint`: List of table-level constraints. A comma is used as a separator.

For complete options, please check out the *CREATE TABLE* documentation by *Cockroach Labs* at the following link: `https://www.cockroachlabs.com/docs/stable/create-table.html`.

Here is an example of the `CREATE TABLE` statement being deployed:

```
CREATE TABLE users (
id INT NOT NULL,
name STRING NOT NULL,
age INT NOT NULL,
PRIMARY KEY (id)
);
```

Optionally, you can also create a **schema**. In the naming hierarchy, a cluster can have multiple databases, a database can have multiple schemas, and a schema can contain multiple tables, views, and sequences. A schema can be created using the `CREATE SCHEMA <schema_name>` statement.

CREATE VIEW

`CREATE VIEW` is used for creating views. A **view** is a virtual table that stores the result of a query. Whenever there is a change in the data in the original table, it is automatically updated. Views don't take up additional physical storage space.

A materialized view is similar to a view but it's physically stored separately. When the data is changed in the original table, materialized views don't get updated automatically. You can use `REFRESH MATERIALIZED VIEW <view_name>` to refresh the contents of a materialized view.

Views are useful for the following reasons:

- They're helpful if you don't want to expose all the columns in the original table due to security concerns.

- Views can contain query results of complex queries, which you don't have to execute explicitly every time.

- A single view can have data from multiple tables and databases.

- They provide meaningful aliases for column names.

- Materialized views can be used for better performance.

CREATE VIEW accepts the following parameters:

- MATERIALIZED: Creates a materialized view.

- IF NOT EXISTS: Creates a view if a view with the same name doesn't already exist.

- OR REPLACE: Creates a view if it doesn't already exist and replaces the view if it already exists. When replacing an existing view, columns in the previous view must appear in the same order as a prefix. However, additional columns are allowed.

- view_name: Name of the view to be created.

- column_name_list: Comma-separated list of column names.

- AS select_statement: A SELECT query, whose results are stored in the view.

Let's look at example of how views are helpful. Let's first insert some records into the users table that was created previously, as follows:

```
INSERT INTO users(id, name, age) values ( 1, 'foo', 13);
INSERT INTO users(id, name, age) values ( 2, 'bar', 24);
INSERT INTO users(id, name, age) values ( 3, 'alice', 14);
INSERT INTO users(id, name, age) values ( 4, 'bob', 29);
> select * from users;
 id | name  | age
----+-------+------
  1 | foo   | 13
  2 | bar   | 24
  3 | alice | 14
  4 | bob   | 29
(4 rows)
```

Now, let's say you want to track all users who are in the age group of 12-17.

You can run this query on the `users` table itself and get the result, as follows:

```
> SELECT * FROM users where age > 12 AND age < 17;
  id | name  | age
-----+-------+------
   1 | foo   | 13
   3 | alice | 14
```

Now, let's say you don't want to run any query directly on the `users` table. Then, you can create a view with that condition and query the view, like this:

```
> CREATE VIEW vaccine_big_kids_group
  AS SELECT id, name, age
  FROM users
  WHERE age > 12 AND age <17;
```

Let's query the view and check the results, as follows:

```
> select * from vaccine_big_kids_group;
  id | name  | age
-----+-------+------
   1 | foo   | 13
   3 | alice | 14
(2 rows)
```

Let's insert one more record that falls under the same age group of 12-17, as follows:

```
INSERT INTO users(id, name, age) values ( 5, 'john', 15);
```

Since the view directly refers to the original table, it's automatically updated, as shown here:

```
> select * from vaccine_big_kids_group;
  id | name  | age
-----+-------+------
   1 | foo   | 13
   3 | alice | 14
   5 | john  | 15
(3 rows)
```

Instead of a view, if you had created a materialized view, you would have had to explicitly refresh in order to get the latest data. Here is an example of creating a materialized view and refreshing it:

```
> CREATE MATERIALIZED VIEW vaccine_big_kids_group_materialized
  AS SELECT *
  FROM users
  WHERE age > 12 AND age <17;
REFRESH MATERIALIZED VIEW vaccine_big_kids_group_materialized;
```

CockroachDB also supports session-scoped temporary views, which are automatically dropped at the end of a session. You can use CREATE VIEW TEMP <view_definition> to create a temporary view.

ALTER

The ALTER statement is used to modify an existing schema object. You can alter the definition of several schema objects such as DATABASE, SCHEMA, TABLE, COLUMN, TYPE, USER, INDEX, VIEW, and so on. We will go over ALTER TABLE and ALTER INDEX in this subsection.

ALTER TABLE takes the following parameters:

- ADD COLUMN: Adds one or more columns
- ADD CONSTRAINT: Adds a constraint to a column
- ALTER COLUMN: Modifies an existing column
- ALTER PRIMARY KEY: Changes the **primary key (PK)**
- DROP COLUMN: Removes a column or multiple columns
- DROP CONSTRAINT: Removes column-level constraints

ALTER INDEX takes the following parameters:

- CONFIGURATION ZONE: Configures replication zones for the index
- RENAME: Renames the index

Here is an example of a column being added:

```
ALTER TABLE users ADD COLUMN new_column INT;
```

```
> SHOW COLUMNS from users;
  column_name | data_type | is_nullable | column_default |
generation_expression |  indices  |  is_hidden
--------------+-----------+-------------+----------------+-----
--------------+-----------+------------
     id       | INT8      |    false    |
NULL          |           |             | {primary} |   false
    name      | STRING    |    false    |
NULL          |           |             | {}        |   false
    age       | INT8      |    false    |
NULL          |           |             | {}        |   false
  new_column  | INT8      |    true     |
NULL          |           |             | {}        |   false
(4 rows)
```

Here is an example of a constraint being added:

```
> ALTER TABLE users ADD CONSTRAINT age_check CHECK (age > 0);
INSERT INTO users(id, name, age) values ( 0, 'cindy', -1);
ERROR: failed to satisfy CHECK constraint (age > 0:::INT8)
SQLSTATE: 23514
CONSTRAINT: age_check
```

DROP

The DROP statement is used to delete a schema object and all the data within it.

The general syntax for DROP is DROP <SCHEMA TYPE> name.

You can see an example here:

```
DROP DATABASE <database_name>;
DROP ROLE <role_name>;
DROP TABLE <table_name>;
```

DROP TABLE takes the following parameters:

- IF EXISTS: Drops the table if it exists; if not, it doesn't return any error
- table_name: Name of the table to drop
- CASCADE: Drops all schema objects that depend on the table, such as views and constraints
- RESTRICT: Restricts the table from getting dropped, if any objects depend on it

Now, let's look at DML statements in the next section.

DML

DML statements are used for managing data within schema objects. They consist of INSERT, UPDATE, UPSERT, and DELETE statements and are generally referred to as statements that insert, update, or delete the data in a database. We will go over the syntax of INSERT, UPDATE, UPSERT, and DELETE statements used for a table in this section.

The INSERT statement takes the following parameters:

- common_table_expression: **Common table expressions (CTEs)** provide a shorthand for a subquery, to improve the readability.
- table_name: Name of the table into which the data is inserted.
- AS table_alias: Alias for the table name.
- column_name: Name of the column that is being populated.
- select_statement: A selection query, whose result is used to insert the data. The column order and column data types of the SELECT query result should match that of the table into which the data is getting inserted.

Here is an example of INSERT INTO:

```
INSERT INTO users(id, name, age) values ( 0, 'cindy', 15);
```

The UPDATE statement updates the existing rows of a table and takes the following parameters:

- common_table_expression: CTEs provide a shorthand for a subquery, to improve the readability.
- table_name: Name of the table in which the rows are updated.
- AS table_alias: Alias for the table name.
- column_name: Name of the column that is being updated.
- a_expression: A new value you want to update, an aggregate function, or a scalar expression used to derive the value.
- FROM table_reference: Specifies a table to reference, but not update.
- select_statement: A selection query, whose result is used to update the data. The column order and column data types of the SELECT query result should match that of the table where the data is getting updated.
- WHERE a_expression: An expression that should evaluate to a Boolean value. A row is updated if this expression returns TRUE.
- sort_clause: An ORDER BY clause.
- limit_clause: A LIMIT clause.

Here is an example of the UPDATE statement being deployed, and also the table data before and after the update:

```
> select * from users;
 id |  name  | age | new_column
----+--------+-----+-------------
  0 | cindy  |  15 |      NULL
  1 | foo    |  13 |      NULL
  2 | bar    |  24 |      NULL
  3 | alice  |  14 |      NULL
  4 | bob    |  29 |      NULL
  5 | john   |  15 |      NULL
(6 rows)

> UPDATE users SET age = 39 WHERE id = 4;
> select * from users;
 id |  name  | age | new_column
```

```
-----+-------+-----+-------------
   0 | cindy |  15 |        NULL
   1 | foo   |  13 |        NULL
   2 | bar   |  24 |        NULL
   3 | alice |  14 |        NULL
   4 | bob   |  39 |        NULL
   5 | john  |  15 |        NULL
(6 rows)
```

The UPSERT statement inserts rows if they don't violate uniqueness constraints, and it updates rows if the values violate uniqueness constraints. UPSERT only looks at PKs for uniqueness.

The UPSERT statement takes the following parameters:

- common_table_expression: CTEs provide a shorthand for a subquery, to improve the readability.

- table_name: Name of the table into which the data is upserted.

- AS table_alias: Alias for the table name.

- column_name: Name of the column that is being populated during the upsert.

- select_statement: A selection query, whose result is used to upsert the data. The column order and column data types of the SELECT query result should match that of the table into which the data is getting upserted.

- DEFAULT VALUES: Used to fill a column with its default value, instead of a SELECT query. Here's an example of this:

```
> select * from users;
  id | name  | age | new_column
-----+-------+-----+-------------
   0 | cindy |  15 |        NULL
   1 | foo   |  13 |        NULL
   2 | bar   |  24 |        NULL
   3 | alice |  14 |        NULL
   4 | bob   |  39 |        NULL
   5 | john  |  15 |        NULL
(6 rows)
> UPSERT INTO users(id, name, age) VALUES (2, 'bar', 34);
> select * from users;
```

```
 id | name  | age | new_column
----+-------+-----+-------------
  0 | cindy | 15  |      NULL
  1 | foo   | 13  |      NULL
  2 | bar   | 34  |      NULL
  3 | alice | 14  |      NULL
  4 | bob   | 39  |      NULL
  5 | john  | 15  |      NULL
(6 rows)
```

The DELETE statement deletes rows from a table. It takes the following parameters:

- common_table_expression: CTEs provide a shorthand for a subquery, to improve the readability.

- table_name: Name of the table in which the data is deleted.

- AS table_alias: Alias for the table name.

- WHERE a_expression: An expression that should evaluate to a Boolean value. A row is updated if this expression returns TRUE.

- sort_clause: An ORDER BY clause.

- limit_clause: A LIMIT clause.

Have a look at the following example:

```
> select * from users;
 id | name  | age | new_column
----+-------+-----+-------------
  0 | cindy | 15  |      NULL
  1 | foo   | 13  |      NULL
  2 | bar   | 34  |      NULL
  3 | alice | 14  |      NULL
  4 | bob   | 39  |      NULL
  5 | john  | 15  |      NULL
(6 rows)

> DELETE from users where id = 4;
> select * from users;
 id | name  | age | new_column
```

```
- - - - - + - - - - - - - + - - - - - + - - - - - - - - - - - - -
     0 |  cindy  |  15  |        NULL
     1 |  foo    |  13  |        NULL
     2 |  bar    |  34  |        NULL
     3 |  alice  |  14  |        NULL
     5 |  john   |  15  |        NULL
(5 rows)
```

In the next section, we will go through some examples of DQL.

DQL

DQL is used for reading or querying data or table metadata. SQL statements involving SELECT and SHOW fall under this category.

SELECT is a very commonly used SQL syntax to read table data. It takes the following parameters:

- ALL: Doesn't eliminate duplicate rows.
- DISTINCT: Eliminates duplicate rows.
- DISTINCT ON (a_expression): Eliminates duplicate rows based on a scalar expression.
- target_element: A scalar expression to determine a column in each result row, or to retrieve all columns in case of an asterisk (*).
- table_expression: A table expression from which the data has to be retrieved.
- AS OF SYSTEM TIME   timestamp: Retrieves data as it existed at the time of this timestamp, where timestamp refers to a specific time in the past. This can return historical data, which can be stale.
- WHERE a_expression: Only retrieves rows that return TRUE for a_expression.
- GROUP BY a_expression: Groups results on one or more columns.
- HAVING a_expression: Only retrieves aggregate function groups that return TRUE for a_expression.
- WINDOW window_definition_list: List of window definitions. The window function computes values by operating on one or more rows returned by a SELECT query.

`SHOW SQL` syntax is commonly used to retrieve the table metadata.

You can use the following queries to list down databases, tables in a database, and columns for a given table respectively:

```
SHOW DATABASES;
SHOW TABLES;
SHOW COLUMNS FROM <table_name>;
```

`SHOW STATISTICS FOR TABLE <table_name>;` is useful for looking at table statistics. This data is used by the cost-based optimizer to improve query performance. This also comes in handy when debugging slow queries.

`SHOW` can be combined with keywords such as `COLUMNS`, `DATABASES`, `TABLES`, `CREATE`, `RANGES`, `STATISTICS`, `TYPES`, `USERS`, `ROLES`, `REGIONS`, and so on to retrieve the metadata associated with a given schema object.

Next, we will go over the list of data types supported by CockroachDB.

Supported data types

In this section, we will go through a list of data types that CockroachDB supports. Here is a list of data types and their descriptions:

- `ARRAY`: Single-dimensional, homogenous array of any non-array data type—for example, `{"cockroachdb", "spanner", "yugabytedb"}`.
- `BIT`: String of binary digits.
- `BOOL`: Boolean value.
- `BYTES`: String of binary characters.
- `DATE`: Date.
- `ENUM`: User-defined data type that consists of a list of static values—for example, `ENUM('earth', 'mars', 'venus')`.
- `DECIMAL`: Exact, fixed-point number.
- `FLOAT`: A 64-bit, inexact, floating-point number.
- `INET`: **Internet Protocol version 4 (IPv4)** and **version 6 (IPv6)** address.
- `INT`: Signed integer.
- `INTERVAL`: Span of time—for example, `INTERVAL '5h39m23s'`.

- JSONB: **JavaScript Object Notation (JSON)** data. Here is an example of JSON data:

```
'{
    "id": 12345,
    "name": "elan",
    "isAlien": "true"
}'
```

- SERIAL: Pseudo data type that combines an integer with a DEFAULT expression. A DEFAULT expression generates different values every time it is evaluated. It ensures a given column gets a value if the INSERT statement doesn't specify a value for it, instead of populating it as NULL.

- STRING: String of Unicode characters.

- TIME: Time in **Coordinated Universal Time (UTC)**.

- TIMETZ: Time value with a specified time zone offset from UTC.

- TIMESTAMP: Stores time and date in UTC.

- TIMESTAMPTZ: Time and date, with a specified time zone offset from UTC.

- UUID: 128-bit hexadecimal value.

In the next section, we will learn about database constraints and various types of constraints in CockroachDB.

Column-level constraints

Constraints are rules that are enforced on data columns in a table. Whenever there is any change in data within a table, all the constraints are verified to make sure none is violated. If violated, the changes are rejected with the appropriate error message. Here is a list of column-level constraints:

- CHECK <condition>: A given condition is checked whenever a value is being inserted into the table. The condition is a Boolean expression that should evaluate to TRUE or NULL. If it returns FALSE for any value, the entire statement is rejected. It is possible to have multiple checks for the same column.

  Here's an example to illustrate this:

```
CREATE TABLE user (
id INT NOT NULL,
name STRING NOT NULL,
```

```
age INT NOT NULL CHECK (age > 18) CHECK (age < 65),
PRIMARY KEY (id)
);
```

- DEFAULT: The DEFAULT value constraint is exercised whenever the INSERT statement doesn't explicitly insert a specific value or NULL for the column that has a DEFAULT constraint. A data type of a DEFAULT value should be the same as that of the original column.

Here's an example to illustrate this:

```
CREATE TABLE employees (
id INT NOT NULL,
name STRING NOT NULL,
salary FLOAT NOT NULL,
bonus FLOAT DEFAULT 0.0,
PRIMARY KEY (id)
);
```

Now, if you insert a row without providing any value for the bonus column (for example, INSERT INTO employee (id, name, salary) VALUES (1, 'foo', 10000.00);), CockroachDB inserts the default value 0.0 for the bonus column.

- FOREIGN KEY: This refers to columns in some other table.

Let's look at an example whereby we create an employees table, as shown in the following code snippet, followed by an employee_info table in which the emp_id column refers to the id column of the employees table:

```
CREATE TABLE employees (
id INT NOT NULL,
name STRING NOT NULL,
salary FLOAT NOT NULL,
bonus FLOAT DEFAULT 0.0,
PRIMARY KEY (id)
);
```

Here is an example where the emp_id column references the id column in the previous employees table:

```
CREATE TABLE employees_info (
emp_id INT NOT NULL,
```

```
address STRING NOT NULL,
CONSTRAINT fk_emp_id FOREIGN KEY (emp_id)   REFERENCES
employees (id)
);
```

Alternatively, an `employees_info` table can also be created, as follows:

```
CREATE TABLE employees_info (
emp_id INT NOT NULL REFERENCES employees(id),
address STRING NOT NULL
);
```

- NOT NULL: Use this if you don't want a column to have NULL values. Any INSERT or UPDATE statement that tries to insert a NULL value will be rejected.

- PRIMARY KEY: The PRIMARY KEY constraint specifies that a column or set of columns that are part of a PK must uniquely identify each row in a table. Tables can have only one PK.

Here's an example to illustrate this:

```
CREATE TABLE employees (
id INT NOT NULL,
name STRING NOT NULL,
salary FLOAT NOT NULL,
bonus FLOAT DEFAULT 0.0,
PRIMARY KEY (id)
);
```

- UNIQUE: The UNIQUE constraint, shown in the following code snippet, ensures that any non-NULL column has a unique value:

```
CREATE TABLE user (
id INT NOT NULL,
email STRING UNIQUE,
name STRING NOT NULL,
age INT NOT NULL CHECK (age > 18) CHECK (age < 65),
PRIMARY KEY (id)
);
```

In the next section, we will go over various types of joins.

Table joins

Table joins are used to combine data from more than one table based on certain conditions on a certain column or columns.

For example, let's assume the following two tables exist:

```
CREATE TABLE customers (
    id UUID PRIMARY KEY,
    name STRING NOT NULL
);
CREATE TABLE purchase_orders (
    id UUID PRIMARY KEY,
    customer_id UUID,
    n_of_items INT,
    total_price DECIMAL(10,2)
);
```

Now, let's look at each of the JOIN types with an example, as follows:

- INNER JOIN: Returns rows from the left and right operands that match the condition.

  Let's look at the following example involving an inner join between the customers and purchase_orders tables:

  ```
  SELECT a.id as customer_id, a.name AS customer_name, b.id
  AS purchase_order_id  FROM customers AS a
  INNER JOIN purchase_orders AS b ON a.id = b.customer_id;
  ```

  This returns all customers and their purchase orders that have matching customers.id values with that of purchase_orders.customer_id.

- LEFT OUTER JOIN: All values from the left table, and for every left row where there is no match on the right, NULL values are returned for the columns on the right.

  Let's look at the following example involving a left outer join between the customers and purchase_orders tables:

  ```
  SELECT a.id as customer_id, a.name AS customer_name, b.id
  AS purchase_order_id  FROM customers AS a
  LEFT OUTER JOIN purchase_orders AS b ON a.id =
  b.customer_id;
  ```

The preceding query will return all customers, whether they have any purchase orders or not.

- RIGHT OUTER JOIN: All values from the right table, and for every right row where there is no match on the left, NULL values are returned for the columns on the left.

  Let's look at the following example involving a right outer join between the customers and purchase_orders tables:

  ```
  SELECT a.id as customer_id, a.name AS customer_name, b.id
  AS purchase_order_id  FROM customers AS a
  RIGHT OUTER JOIN purchase_orders AS b ON a.id =
  b.customer_id;
  ```

  The preceding query will return all purchase_orders values, whether they have a corresponding customer in the customers table or not.

- FULL JOIN: For every row on one side of the join where there is no match on the other side, NULL values are returned for the columns on the non-matching side.

  Have a look at the following example to illustrate this:

  ```
  SELECT a.id as customer_id, a.name AS customer_name, b.id
  AS purchase_order_id  FROM customers AS a
  FULL OUTER JOIN purchase_orders AS b ON a.id =
  b.customer_id;
  ```

  The preceding query will return all the rows from both tables.

Next, we will learn about sequences and the syntax for creating a sequence.

Using sequences

Sequences are helpful when you need an auto-increment integer sequence in a table.

Let's look at the following example in which we create a sequence with the default setting:

```
root@localhost:26258/test> CREATE SEQUENCE customer_id_seq;
CREATE SEQUENCE


Time: 158ms total (execution 158ms / network 0ms)


root@localhost:26258/test> SHOW CREATE customer_id_seq;
     table_
name      |                                              create_
```

```
statement
------------------+-----------------------------------------------------
-----------------------------------------------------------
  customer_id_seq | CREATE SEQUENCE public.customer_id_seq
MINVALUE 1 MAXVALUE 9223372036854775807 INCREMENT 1 START 1
(1 row)


Time: 136ms total (execution 135ms / network 1ms)
```

Here is an example of using the previously created sequence in a table column:

```
CREATE TABLE customers (
    id INT,
    row_id INT DEFAULT nextval('customer_id_seq'),
    name STRING
);
```

Here, whenever a row is inserted, DEFAULT will call nextval, which will generate increments in the customer_id_seq sequence and use that as the value for the row_id column.

Next, we will go through the benefits of online schema changes.

Managing schema changes

CockroachDB supports online schema changes, which don't require any downtime. An existing schema can be changed using statements that include ALTER and CREATE INDEX operations.

CockroachDB internally maintains a consistent distributed schema cache, along with consistent table data that works with multiple versions of the schema concurrently. This enables the rolling out of a new schema while the older schema is still being used. CockroachDB backfills the table data into the newer schema without holding locks. So, online schema changes don't affect the current read/write operations on the cluster on a particular table whose schema is being modified.

Here are some of the benefits of online schema changes:

- Zero downtime.

- Schema changes happen without holding any table-level locks, so application workloads on the cluster can continue to operate without performance degradation.

- Data is kept consistent throughout the schema upgrade.

Here are some of the known limitations of schema changes:

- If you are using a multi-statement transaction in which you combine DDL and non-DDL statements, DDL statements can fail at COMMIT, while non-DDL statements might have been committed. This can result in an inconsistent state, whereby some of the statements in a transaction are committed and some of them are aborted. It's important to look for appropriate error codes and handle them correctly.

- You have to pay extra attention if you are using prepared statements on a table whose schema is changed before the prepared statement is executed. This can result in inconsistencies.

Please refer to a complete list of known limitations here: https://www.cockroachlabs.com/docs/stable/known-limitations.html.

Some of these limitations might get fixed in future releases, and there might also be new ones.

With that, we have reached the end of this chapter.

Summary

In this chapter, we learned about the SQL syntax for DDLs, DMLs, and DQLs. We also went over other popular features such as indexes, joins, and sequences. We have left out some of the experimental and enterprise-only features. Also, whenever you are not sure about the syntax of a given SQL operation, please use /h to get detailed information about the SQL operation.

In the next chapter, we will go over the admin **user interface** (**UI**) and learn how to use it to manage a CockroachDB cluster.

8
Exploring the Admin User Interface

In *Chapter 7, Schema Creation and Management*, we learned about various SQL syntaxes and how to manage schema changes. In this chapter, we will learn what information is available in the admin **user interface** (**UI**) for monitoring a CockroachDB cluster.

The admin UI is useful in monitoring the health of a CockroachDB cluster as it provides the statistics, metrics, and status of all the nodes in the cluster. The admin UI mainly provides information about the metrics, database and table definitions, sessions, transactions, network latencies, active jobs, and advanced debugging information. **Network latency** is important when debugging query latencies, just to rule out slowness due to the network. Also, if you have multi-cloud or hybrid cloud CockroachDB clusters, it is important to understand the latencies across various cloud providers. Schema changes, backup and restore jobs, data imports, and changefeed are covered on the **Jobs** tab.

Network latency and advanced debugging will be covered in *Chapter 10, Troubleshooting Issues*.

The following topics will be covered in this chapter:

- Introducing the admin UI
- Cluster overview
- Metrics deep dive
- Database and table definitions
- Understanding sessions
- Transactions
- Tracking jobs

Technical requirements

Although we are not going to execute any examples in this chapter, we still need a cluster to access the admin UI. If you still haven't installed a CockroachDB cluster, please refer to the *Technical requirements* section in *Chapter 2, How Does CockroachDB Work Internally?*

We will first start with a general introduction on how to access the UI and how to read the metrics.

Introducing the admin UI

The CockroachDB admin UI comes by default, and you don't need any special setup to configure one. The admin UI can be accessed from any node within the cluster using the IP address or the hostname of the node and the port that's configured using the `--http-addr` flag, for example, `--http-addr=localhost:8080`. If this flag is not configured, then the admin UI is available through the IP address or hostname specified in `--listen-addr` and the default port `8080`.

The cluster overview is the landing page in the UI, as shown in the following screenshot:

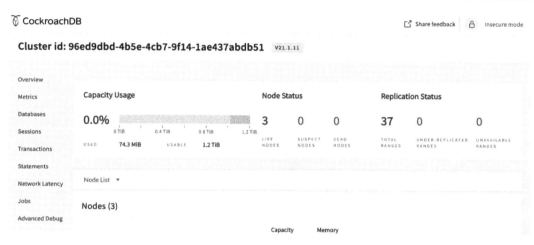

Figure 8.1 – Landing page in the UI – Cluster overview

With regard to the metrics, all the information in the UI can be viewed at either the individual node level or at the cluster level. You can see that there is a dropdown in the UI that lets you select individual nodes or the cluster itself, as shown in the following screenshot. In the case of a cluster, the stats and metrics are averaged down or aggregated across all the participating nodes in the cluster. You can look at the top-right corner to find out whether the cluster is running in insecure mode.

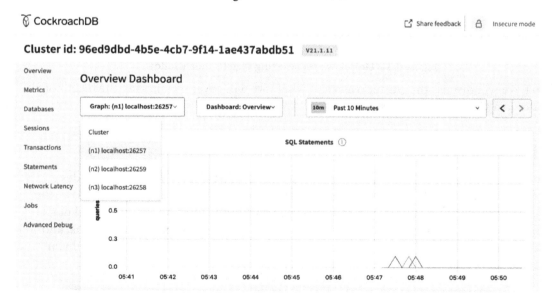

Figure 8.2 – Dropdown for selecting individual nodes for the node view or cluster view

Also, the UI lets you customize the time window, as shown in the following screenshot:

Figure 8.3 – Dropdown for selecting various time windows

Next, we will discuss the cluster overview, where we will cover monitoring the health of all nodes in a CockroachDB cluster.

Cluster overview

This is the page you see when you open the admin UI. The cluster overview provides a high-level overview of the CockroachDB cluster. *Figure 8.4* is a screenshot that shows the cluster overview for a cluster with three nodes. Some of the links on the side panel don't show up if there is no data. For example, you might not see **Sessions**, **Transactions**, and **Statements** until a session has been established with the CockroachDB cluster and you execute some queries.

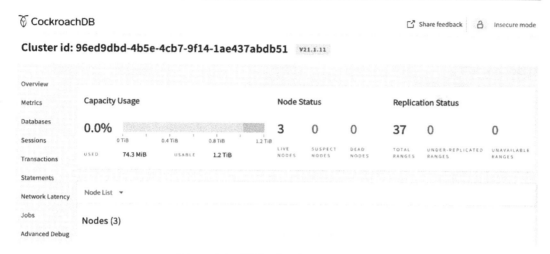

Figure 8.4 – UI for the cluster overview

You can find the following set of information in the cluster's **Overview** tab:

- **Capacity Usage**: This is the total disk space used by all the nodes in the CockroachDB cluster. It classifies the usage further into **USED** and **USABLE**. **USED** indicates the current disk space by all the nodes, and **USABLE** is the maximum capacity that can eventually be used by all the nodes in the cluster. You can set the maximum capacity for a given node by using the `--store` option. So, if the `--store` option is provided, the total usage space cannot exceed this configured value.

- **Node Status**: This shows the number of live, dead, and suspect nodes. A live node is alive and healthy. If a node is shut down, it is considered dead. A suspect status indicates that either the node cannot be reached to ascertain its status or that the node is getting decommissioned.

- **Replication Status**: This shows information about the total number of ranges in the cluster. It also shows under-replicated and unavailable ranges. A range would be under-replicated if the number of replicas of that range is less than the configured replication factor. If the majority of replicas of a given range are unavailable, then that range itself becomes unavailable as you cannot have consensus. If a range is unavailable, all the queries involving that range will fail.

The following screenshot shows the node list, where you can see some details about each of the nodes in that cluster:

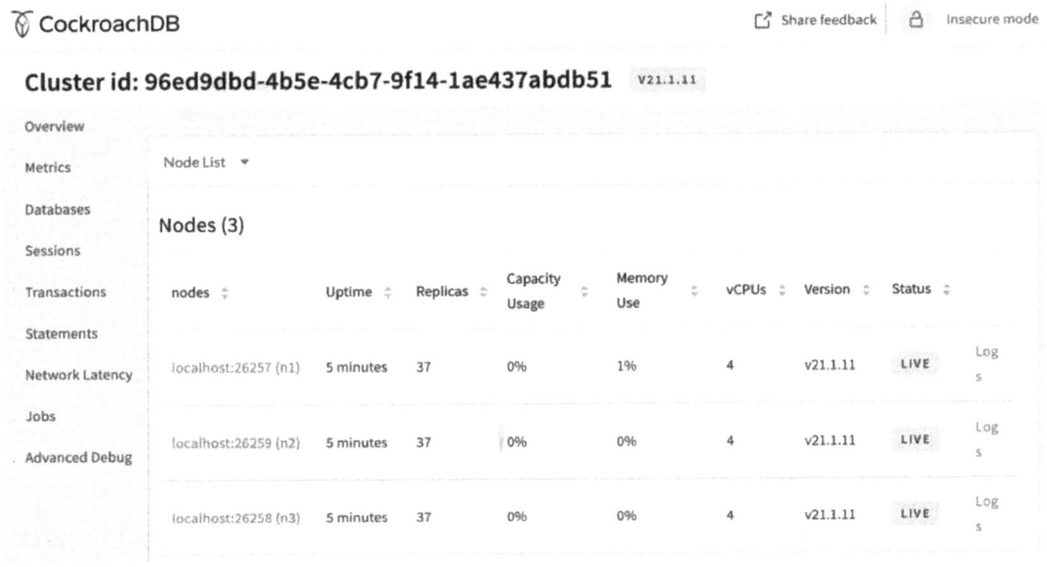

Figure 8.5 – Cluster overview showing the node list

Node List provides the list of nodes with individual node-level information. Node-level information includes the following:

- **Region/Node Address**: If you have configured different regions, then the nodes are grouped by regions. Under each region, you will see the IP address and port of the nodes in that region. If the regions are not configured, you would just see the node IP address and the port.

- **Uptime**: The duration for which the node is running.

- **Replicas**: The number of replicas.

- **Capacity Usage**: How much of the disk space has been used up.

- **Memory Use**: How much of the memory has been used up.

- **vCPUs**: The number of CPUs.

- **Version**: The version of CockroachDB running on this node.

- **Logs**: The logs relevant to all the operations on that node. Logs are useful for debugging issues and for ascertaining whether the node is healthy.

- **Status**: This tells us the health of a node. **LIVE** means the node is healthy and serving traffic, while **DEAD** means the node is down and is no longer serving traffic.

In the next section, we will go through the various types of metrics and how they can be useful in understanding query latencies.

Metrics deep dive

Metrics are very useful for measuring the general health of individual nodes and provide better insights into the overall cluster when we are debugging performance-related issues. Metrics can also be filtered based on different time windows.

The Metrics dashboard comes with a lot of useful metrics that are shown in the following screenshot:

Figure 8.6 – Metrics dashboard showing various options in the dropdown

The Metrics dashboard includes the following categories:

- **Hardware**: Here, you can find the following pieces of information:

  - **CPU Percent**: The percentage of the CPU being consumed by the CockroachDB process

- **Memory usage**: The memory used by the CockroachDB process
- **Disk Read Mebibytes / second**: The average number of bytes read from the disk by all the processes, expressed in Mebibytes per second
- **Disk Write Mebibytes / second**: The average number of bytes written to the disk by all the processes, expressed in Mebibytes per second
- **Disk Read IOPS**: The number of disk read operations by all the processes per second, averaged for 10 seconds
- **Disk Write IOPS**: The number of disk write operations by all the processes per second, averaged for 10 seconds
- **Disk Ops In Progress**: The number of read and write operations in the queue from all processes
- **Available Disk Capacity**: Available storage capacity
- **Network Bytes Received**: The number of bytes received over the network per second for all processes, averaged for 10 seconds
- **Network Bytes Sent**: The number of bytes sent over the network per second for all processes, averaged for 10 seconds

- **Runtime:**

  - **Live Node Count**: Number of active nodes in the cluster
  - **Memory Usage**: Detailed memory usage, which is broken down further into the following items:

    - **Resident Set Size (RSS)**: RSS indicates the size of the subset of the memory occupied by a process in the **RAM** (short for **Random Access Memory**). This doesn't include the swap memory. **Swap memory** is a portion of the process's runtime memory that is swapped out into the disk. In the context of CockroachDB, RSS indicates the total memory used by the CockroachDB process.
    - **Go Allocated**: Memory allocated by GoLang.
    - **Go Total**: Total memory managed by GoLang.
    - **CGo Allocated**: Memory allocated by C. CGo enables Go packages to call C code.
    - **CGo Total**: Total memory managed by C.

> **Note**
>
> CGo enables GoLang packages to call C code as it was originally required when CockroachDB was using RocksDB as its storage engine, which was primarily written in C and C++. CGo usage might have reduced after CockroachDB moved to Pebble. Pebble is written in GoLang.

- **Goroutine Count**: Number of Go routines.

- **Runnable Goroutines per CPU**: Number of Goroutines waiting for the CPU.

- **GC Runs**: Number of times GoLang's garbage collector has run per second.

- **GC Pause Time**: CPU time per second used by GoLang's garbage collector. During garbage collection, CockroachDB's regular database operations are paused.

- **CPU Time**: CPU time used by the CockroachDB process for user- and system-level operations.

- **Clock Offset**: In node view, it shows the mean clock offset of a given node against the rest of the nodes in the cluster. In cluster view, it shows the mean clock offset of each node against other nodes in the cluster.

- **SQL:**

  - **Open SQL Sessions**: In node view, this is the number of SQL connections open from CockroachDB client(s) to a given CockroachDB node. In cluster view, this is the total number of SQL connections open from CockroachDB client(s) to all the CockroachDB nodes in the cluster. In node view, this is the number of bytes/second transferred between the clients and the node.

  - **Open SQL Transactions**: Number of open SQL transactions.

  - **Active SQL Statements**: Number of SQL statements that are currently getting executed.

  - **SQL Byte Traffic**: SQL client network traffic in bytes per second.

  - **SQL Statements**: 10-second average of the `SELECT`, `INSERT`, `UPDATE`, and `DELETE` statements.

  - **SQL Statement Errors**: Number of SQL statements that resulted in an error.

  - **SQL Statement Contention**: Number of SQL statements that experienced contention.

  - **Active Flows for Distributed SQL Elements**: Number of flows on each node in the cluster that are participating in active, distributed SQL statements.

- **Service Latency**: SQL, 99th and 90th percentiles: Within the last minute, 99% and 90% of the queries are executed within this time, respectively.

- **KV Execution Latency**: 99th and 90th percentiles: This is the latency between the query request and response times, for 99% and 90% of the queries, respectively.

- **Transactions**: Total number of transactions per second.

- **Transaction Latency**: 99th and 90th percentiles: Total transaction time within the last minute, for 99% and 90% of the transactions, respectively.

- **SQL Memory**: Current allocated SQL memory.

- **Schema Changes**: Total number of **Data Definition Language (DDL)** statements per second.

- **Statement Denials**: Cluster settings: Total number of statements that were rejected due to a cluster setting.

- **Storage:**

  - **Capacity**: Storage capacity is further classified into the following categories:

    - **Max**: Maximum storage size

    - **Available**: Free storage space available

    - **Used**: Storage space currently used

  - **Live Bytes:**

    - **Live**: Number of logical bytes stored in active key-value pairs

    - **System**: Number of physical bytes stored in system key-value pairs

  - **Log Commit Latency**: 99th and 50th percentiles: Raft log commit latency in the 99th and 50th percentiles

  - **Command Commit Latency**: 99th and 50th percentiles: Latency of the raft commit commands in the 99th and 50th percentiles

  - **Read Amplification**: The average number of real read operations executed per logical read operation across all nodes

  - **SSTables**: Number of **Sorted Strings Tables (SSTable)** in use

  - **File Descriptors**: Number of open file descriptors

  - **Compactions/Flushes**: Number of compaction and memtable flushes per second.

- **Time Series Writes**: Number of time series writes per second, including the attempts that errored out
- **Time Series Bytes Written**: Number of bytes written by time series per second

- **Replication:**

  - **Ranges**: This provides the following information:

    - **Ranges**: Total number of ranges.
    - **Leaders**: Number of leaders.
    - **Lease Holders**: Number of lease holders.
    - **Leaders w/o lease**: Exclusive raft leaders that are not lease holders.
    - **Unavailable**: Unavailable ranges due to the majority of replicas being unavailable.
    - **Under-replicated**: Under-replicated ranges that are replicated less than the replication factor.
    - **Over-replicated**: Over-replicated ranges. This usually happens when more nodes are added.

  - **Replicas Per Node**: Number of replicas on each node.
  - **Leaseholders Per Node**: Number of leaseholders per node.
  - **Average Queries Per Node**: Exponentially weighted moving average of the number of KV batch requests processed by leaseholder replicas on each node per second. Tracks roughly the last 30 minutes of requests. Used for load-based rebalancing decisions.
  - **Logical Bytes Per Node**: Number of logical bytes stored in key-value pairs on each node.
  - **Replica Quiescence**:

    - **Replicas**: Number of replicas
    - **Quiescent**: Number of replicas that have not been accessed recently

  - **Range Operations**:

    - **Splits**: Number of range splits.
    - **Merges**: Number of range merges.
    - **Adds**: Number of newly added ranges.

- **Removes**: Number of ranges removed.

- **Lease Transfers**: Number of transfers of a lease for a given range between nodes.

- **Load-based Lease Transfers**: Number of transfers of a lease, to ensure even load balancing across all the nodes.

- **Load-based Range Rebalances**: Number of range rebalances, to ensure all nodes get equal traffic. Can happen when there are **hot ranges**. A range is considered hot if it's getting more read/write requests compared to other ranges.

- **Snapshots**: Snapshots are used when some of the nodes in a Raft group are lagging considerably. In such cases, instead of sending individual messages to nodes that are lagging, a cluster can send a snapshot of the range, which can be directly applied locally. The following are some of the snapshots metrics that are available in the UI:

  - **Generated**: Number of snapshots generated per second

  - **Applied (Voters)**: Number of snapshots applied to nodes per second that were initiated by Raft

  - **Applied (Initial Upreplication)**: Number of snapshots applied to newly joining nodes in order to bring it up to speed

  - **Applied (Non-voters)**: Number of snapshots applied to lagging nodes identified by the cluster

  - **Reserved**: Number of slots reserved per second for incoming snapshots that will be sent to a lagging node

The next three dashboards contain metrics mostly internal to CockroachDB. You can look into them if you are familiar with the internal architecture and code of CockroachDB.

- **Distributed**: This dashboard shows the latency of various types of distributed transactions within CockroachDB and includes the following set of metrics:

  - **Batches**

  - **RPCs**

  - **RPC Errors**

  - **KV Transactions**

  - **KV Transaction Restarts**

- KV Transaction Durations
- Node Heartbeat Latency

- **Queues**: This dashboard provides metrics related to different types of queues used within CockroachDB. You can see the following set of metrics related to different queues:

  - **Queue Processing Failures**

  - **Queue Processing Times**

  - **Replica GC Queue**

  - **Replica Queue**

  - **Split Queue**

  - **Merge Queue**

  - **GC Queue**

  - **Raft Log Queue**

  - **Raft Snapshot Queue**

  - **Consistency Checker Queue**

  - **Time Series Maintenance Queue**

- **Slow Requests**: This dashboard shows metrics related to various internal activities that are being performed slower than expected. The following set of metrics are included as part of slow requests:

  - **Slow Raft Proposals**

  - **Slow DistSender RPCs**

  - **Slow Lease Acquisitions**

  - **Slow Latch Acquisitions**

- **Changefeed**: **Change data capture** (**CDC**) provides a mechanism for tracking all the row-level changes in CockroachDB and feeding the changes to a configurable sink. This is useful for external replication, reporting, caching, and searching. The following are some of the metrics related to changefeed:

  - **Max Changefeed Latency**: Maximum latency for resolved timestamps of any running changefeed

  - **Sink Byte Traffic**: Number of bytes emitted by CockroachDB into the sink

- **Sink Counts**:

  - **Messages**: Number of messages sent by CockroachDB to the sink

  - **Flushes**: Number of flushes that the sink has done for changefeeds

- **Sink Timings**:

  - **Message Emit Time**: The time in milliseconds required by CockroachDB to send the messages to the sink

  - **Flush Time**: The time CockroachDB spent waiting for the sink to flush the message

- **Changefeed Restarts**: The number of changefeed restarts due to retryable errors.

In the next section, we will go over the Databases dashboard.

Database and table definitions

The Databases dashboard contains the list of databases, with a list of tables in them. You can also view grants given to various users.

In the following screenshot, you can see the **Databases** dashboard, with options to view **Tables** and **Grants**:

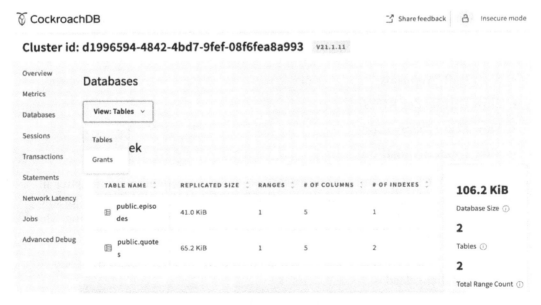

Figure 8.7 – Databases dashboard showing the tables and grants options in the dropdown

If you select the **Tables** option, you will see a list of tables for all the databases with some stats. If the stats are not loaded, you can click on the **Load stats for all tables** option to populate the table-level stats, as can be seen in the following screenshot:

Figure 8.8 – Databases dashboard showing tables

From this dashboard, you can also view the DDLs for all the tables, as shown in the following screenshot:

Figure 8.9 – Databases dashboard showing the DDLs of tables

Grants will show user-level permissions with respect to a given database, as shown in the following screenshot:

Figure 8.10 – Databases dashboard showing user grants

The Databases dashboard is useful for quickly going through the tables and grants, instead of using a SQL client and executing some queries.

In the next section, we will go over sessions.

Understanding sessions

The **Sessions** dashboard gives information about all the active client sessions within the CockroachDB cluster. In the following screenshot, you can see the **Sessions** dashboard showing an active session:

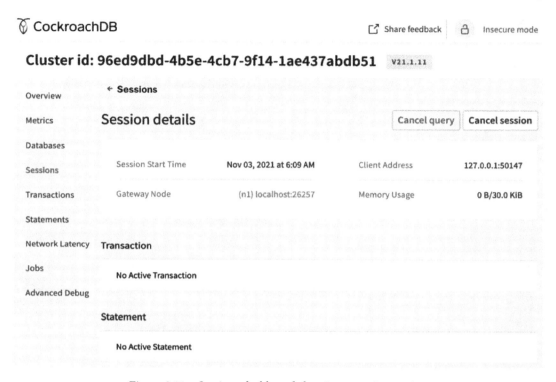

Figure 8.11 – Sessions dashboard showing an active session

The **Sessions** dashboard displays the following information:

- **Session Duration**: Amount of time for which the session is open

- **Transaction Duration**: Amount of time of the current active transaction, if any

- **Statement Duration**: Amount of time of the current, active SQL statement, if any

- **Memory Usage**: The current allocated memory for this session/maximum memory allocated during this session

- **Statement**: The SQL statement that's currently active

- **Actions**: Options to end an active query that is part of this session:

  - **Terminate Statement**: Ends the SQL statement

  - **Terminate Session**: Ends the session

In the next section, we will go over the Transactions dashboard.

Transactions

The **Transactions** dashboard lists all the current transactions and provides additional information, including transaction time, contention, and retries, which help identify slow transactions. The following screenshot is of a **Transactions** dashboard showing transaction details:

Figure 8.12 – Transactions dashboard showing transaction details

The **Transaction** dashboard shows the following details:

- **Transactions**: SQL statements that are part of the transaction
- **Execution Count**: Total number of executions for a given transaction
- **Rows Read**: Average number of rows read from disk during the transaction
- **Bytes Read**: Total number of bytes read across all the statements in a transaction
- **Transaction Time**: Average planning and execution time
- **Contention**: Average time a given transaction was in contention with other transactions

- **Max Memory**: Maximum memory used by a given transaction
- **Network**: Amount of data transferred over the network during this transaction
- **Retries**: Total number of retries during this transaction
- **Statements**: Number of SQL statements in the transaction

In the last section, we will learn about the Jobs dashboard.

Tracking jobs

A job can be one of these activities: backups, restores, imports, schema changes, changefeed, statistics creation, and auto-statistics creation. The following screenshot is of a **Jobs** dashboard:

Figure 8.13 – Jobs dashboard

You can filter jobs based on the job type, as shown in the following screenshot:

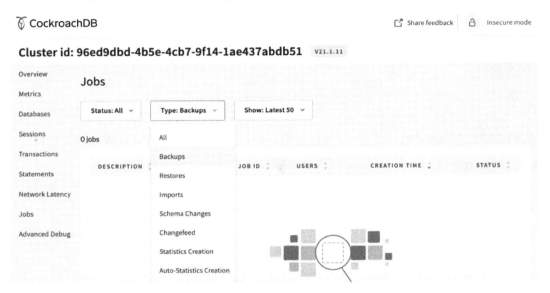

Figure 8.14 – Jobs dashboard showing a dropdown for various job types

You can also filter jobs by their status, as shown in the following screenshot:

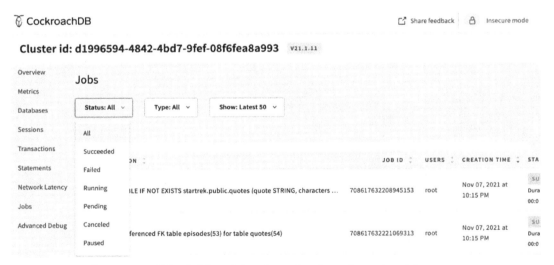

Figure 8.15 – Jobs dashboard showing a dropdown for various job statuses

The following are the various statuses of a job:

- **PENDING**: Job has been created, but not yet started
- **PAUSED**: Job is paused
- **FAILED**: Job has failed during execution
- **SUCCEEDED**: Job has completed successfully
- **CANCELED**: Job was canceled before it could complete

The following information is available for each job:

- **Description**: SQL statement
- **Job ID**: Unique ID assigned to a job for tracking purposes
- **Users**: User who created the job
- **Creation Time**: Date and time when the job was created
- **Status**: Current status of the job

We looked at several metrics that are available on the admin user interface. Although it can be overwhelming to understand and observe all of them, you should at least start with a cluster overview and the metrics dashboard and move on to others once you become an advanced user of CockroachDB.

Summary

In this chapter, we learned about the admin user interface, various dashboards, and all the information available for us to better understand the health of a cluster and also the query latencies. It is very important to get yourself familiarized with the admin UI so that you can easily navigate to the right places when debugging issues. We did not cover network latency and advanced debugging dashboards as these will be covered in detail in *Chapter 10, Troubleshooting Issues*.

In the next chapter, we will cover security.

9

An Overview Of Security Aspects

In *Chapter 8, Exploring the Admin User Interface*, we learned about the admin user interface, which provides several features to help manage the cluster. The admin user interface is also very useful in debugging issues as it provides metrics about the health of the cluster, transactions, and queries.

In this chapter, we will learn about security. The increase in cloud usage has given rise to an enormous amount of surface attacks. Now, hackers have various ways to compromise a given system and leak its data. Ransomware, phishing, malware, spyware, and threatware are some of the most popular programs that intend to cause harm to your systems and leak their data, either to gain popularity or money. Cloud cybersecurity experts are the most sought-after in today's world. Since a transactionally distributed SQL database will be at the heart of any core infrastructure, it's of the utmost importance to pay attention to the security aspects.

In this chapter, we will start by providing a brief introduction to various aspects of security, followed by a deeper discussion of each of those aspects and various options that are available when using CockroachDB. To do this, we will cover the following topics:

- Introduction to security concepts
- Client and node authentication
- Authorization mechanisms

- Data encryption at rest and in flight
- Audit logging
- RTO and RPO
- Keeping the network secure
- Security best practices

Technical requirements

The examples in this chapter require you to have CockroachDB installed. If you still haven't done so, please refer to the *Technical requirements* section of *Chapter 2, How Does CockroachDB Work Internally?*.

Introduction to security concepts

Authentication is required for a SQL client that executes queries against a CockroachDB cluster and for nodes in a cluster that communicate with each other. In this section, we will go over some of the available options for client and node authentication.

Authorization is about deciding who can access what resources. In this section, we will discuss users, roles, and configuring privilege access to various schema objects.

Data at rest refers to data when it's stored on a physical storage device. Encrypting the data that's on a storage device renders it unreadable, even when a hacker gets hold of the encrypted data. Data in flight refers to the data that's on-wire when it's being transferred between the client and the CockroachDB cluster or between the nodes in a CockroachDB cluster. It is important to encrypt the data on-wire as it makes it useless when some middleman manages to sniff the data.

Audit logging is a log collection process that keeps track of all the activities that were performed on the data, including the time, the client's device information, the query, the operation, the event, and the user. This information is very useful if a data leak occurs and we want to investigate or ensure our system is compliant with certain standards.

The **recovery time objective (RTO)** and **recovery point objective (RPO)** are two important parameters when it comes to disaster recovery and data protection. RTO refers to how soon we can resume operations, after a given failure, while RPO refers to the maximum allowed time when we can restore the data. We will discuss these terms while providing examples and learn how to prepare for disaster recovery later in this chapter.

With deployments that involve on-premises, multi-cloud, and hybrid cloud environments, network security is key to denying hackers from gaining access to the data. In this section, we will briefly cover some of the options for providing maximum network security.

In the last section, we will list some of the best security practices.

In the next section, we will learn about authentication and some of the options we can use to provide authentication in CockroachDB.

Client and node authentication

Authentication is the process of verifying the identity of a system that is making a request. In the context of CockroachDB, this can be a client executing queries on a CockroachDB cluster or the nodes in a cluster that are talking to each other. Authentication can be achieved by using certificates and keys. Let's look at an example. Let's assume that foo and bar want to talk to each other and that before they start talking, they want to ensure they are talking to each other. First, we must understand the concept of **public-private keys**. Any message that you encrypt with a public key can be decrypted using its corresponding private key. This pair is supposed to be unique in that no other key can be used for decryption. Also, they have to be different. The following diagram shows how public key encryption works:

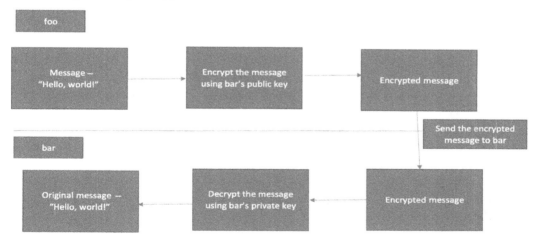

Figure 9.1 – Public key encryption

So, going back to our example, foo and bar have a pair of public and private keys. At the beginning of their communication, their public keys should be exchanged. During this exchange, both parties need a way to know if the public key is coming from an intended recipient, not an imposter, and that the public key is authentic. For example, if foo shares its public key with bar, then bar should have a way to ensure that it's coming from foo and that the public key belongs to foo. This is where a **certificate authority** (**CA**) comes into the picture. The following diagram shows how a digital certificate is generated by the CA once a CSR request is received from the applicant:

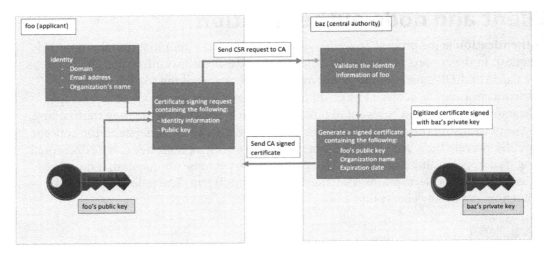

Figure 9.2 – Digital certificate generation by CA

A CA is an entity that issues certificates for other entities. Let's say that baz is a CA. Now, baz can issue a certificate on behalf of foo, which contains foo's public key and some more identity information about foo. This certificate is signed by baz's private key. foo and bar both have the CA's (baz's) public key. So, we now have baz's certificate, which is used to sign foo and bar's certificates. We also have the certificates of foo and bar that have been signed by the CA, which contain their public keys. When the communication between foo and bar starts, they can exchange these certificates and also verify their authenticity as they are issued by the same CA, by using CA's public key. After that, they can securely talk to each other.

First, let's look at some of the CockroachDB commands that can be used to generate certificates and keys.

Generating certificates and keys

The following are some of the options you have to generate certificates and keys that can be used for authentication:

- `create-ca`: Creates the self-signed CA. This can be used to create and authenticate certificates for all the nodes within the cluster. You can also use an external CA for this.

- `create-node`: Creates a certificate and key for a specific node in the cluster. You can specify all the addresses (IP address or hostname) to reach this node.

- `create-client`: Creates a certificate and key for a user accessing the cluster from a client. If you have multiple users, you should generate a separate certificate for each user.

- `list`: Lists the certificates and keys that were found in the CA. These are passed in the input argument.

The following commands can be used to generate a certificate and key for the CA, node, and client:

```
Generate CA certificate.
$ cockroach cert create-ca \
--certs-dir=<certs_directory> \
--ca-key=<CA_key_directory>
```

The following commands can be used to generate the node's certificate and key. Any number of hostnames or addresses can be used to reach the node:

```
$ cockroach cert create-node \
<hostname_1_of_node> \
<hostname_2_of_node> \
--certs-dir=<certs_directory> \
--ca-key=<CA_key_directory>
```

The following commands can be used to generate the client's certificate and key:

```
$ cockroach cert create-client <user_name> \
--certs-dir=<certs_directory> \
--ca-key=<CA_key_directory>
```

The following command lists the certificates and keys in a given directory:

```
$ cockroach cert list \
--certs-dir=<certs_directory>
```

You can also use the **OpenSSL** tool to generate the certificates and keys. This tool can be downloaded from https://www.openssl.org/source/. Please refer to the documentation on the aforementioned website to learn how to use this tool.

Next, we will learn how these certificates and keys can be used to provide client and node authentication.

Client authentication

Whenever a SQL client sends a request to a CockroachDB cluster, it's important to authenticate the client before serving the request. In the following subsections, we will look at some of the ways in which we can ensure client authentication.

Password authentication without TLS

This option is what we have used in all of our examples in this book. It's useful in cases where both the client and server are within a secured perimeter that doesn't need further transport layer security. However, you still need a user to identify yourself. Let's learn how to create a password for a user:

1. First, create a user with a password, as follows:

    ```
    > CREATE USER kishen WITH LOGIN PASSWORD
    'oldmacdonaldhadafarm(I(Io'
    ```

2. Then, you can use that user from the client:

    ```
    $ cockroach sql --user=kishen --insecure
    #
    # Welcome to the CockroachDB SQL shell.
    # All statements must be terminated by a semicolon.
    # To exit, type: \q.
    #
    Enter password:
    ```

Next, we will learn how to perform password authentication using transport layer security.

Password authentication with TLS

If you wish to perform password authentication with **Transport Layer Security** (TLS), you must provide the certificate directory in the input, as follows:

```
$ cockroach sql --user=<user_name> \
  --certs-dir=<certs_directory>
```

Single sign-on (SSO) authentication

Single sign-on (**SSO**) is an Enterprise-only feature that needs an external OAuth 2.0 identity provider. This requires the user to log in to an external identity provider in the admin user interface. Once authenticated, the user is redirected to the CockroachDB cluster through a callback URL. An ID token is used to authorize the callback. This ID token is a security token that contains claims about the authentication of a user by an OAuth 2.0 identity provider. The ID token will be in **JSON Web Token** (**JWT**) form, which is a JSON containing details about the authentication, including the expiry time of the token, the unique subject identifier, and the issuer identifier. CockroachDB can match the ID token to a SQL user and subsequently create a web session for that SQL user. With the web session, the user can access the admin user interface.

Generic Security Services API (GSSAPI) using Kerberos authentication

This is an Enterprise-only feature. You would typically need a Kerberos environment, a GSSAPI-compatible PostgreSQL client, a service principal, and a Kerberos client to be installed to provide this authentication mechanism. Please refer to the GSSAPI setup guide at `https://www.cockroachlabs.com/docs/stable/gssapi_authentication.html` for specific guidelines.

Node authentication

We learned how to generate certificates and keys in the *Generating certificates and keys* section. Once we've generated the node's certificates, we just have to make sure they are configured correctly when we start the node. If they're not, you will see communication errors around the TLS connection when the nodes are trying to talk to each other. You can provide the certificate directory when you start a node like so:

```
$ cockroach start \
  --certs-dir=<certs_directory> \
  --store=node1 \
  --listen-addr=localhost:26257 \
```

```
--http-addr=localhost:8080 \
--join=localhost:26257, localhost:26258, localhost:26259
```

You can also use a tool such as Vault (https://www.vaultproject.io/) to dynamically generate short-lived certificates. This will make them available for all the nodes in a cluster and automatically rotate the certificates.

In the next section, we will learn about authorization and how to give granular permissions to schema objects.

Authorization mechanisms

Authorization involves controlling access to schema objects and giving the minimum required permissions and privileges for users and roles. Authorization becomes critical when the data's size, number of nodes, number of clients, number of clusters, and use cases grow in size.

In the context of CockroachDB, a user and role can be used interchangeably as there is no technical distinction between them. Even when executing CockroachDB commands, role and user can be used interchangeably in some cases. An example of this is as follows:

1. First, we must execute SHOW ROLES:

```
$ SHOW ROLES;
  username | options | member_of
-----------+---------+------------
  admin    |         | {}
  root     |         | {admin}
(2 rows)


Time: 13ms total (execution 12ms / network 1ms)
```

2. Next, we must execute SHOW USERS:

```
$ SHOW USERS;
  username | options | member_of
-----------+---------+------------
  admin    |         | {}
  root     |         | {admin}
```

```
(2 rows)
```

```
Time: 9ms total (execution 9ms / network 0ms)
```

As you can see, both the preceding outputs are identical. Henceforth, we will only use the term *user* for simplicity.

A user can be allowed to perform specific actions on specific schema objects. Also, if a user is a member of another user, all their privileges will be inherited.

A user can be created with the CREATE USER command, as shown in the following code:

```
CREATE USER 'kishen' WITH LOGIN PASSWORD
'oldmacdonaldhadaform(I(I0' VALID UNTIL '2021-12-25';
```

Make sure that you start your cluster by configuring the certificate's directory, without the insecure flag. Otherwise, you will see the following error message:

```
ERROR: setting or updating a password is not supported in
insecure mode
SQLSTATE: 28P01
```

Next, let's look at giving privileges to specific objects.

Roles

You can create roles that perform specific functions. For example, you can create a role for just creating other roles, like so:

```
$ CREATE ROLE can_create_role WITH CREATEROLE CREATELOGIN;
```

When you run SHOW ROLES, the new role should appear:

```
$ SHOW ROLES;
     username     |                 options                 | member_
of
------------------+-----------------------------------------+--------
----
  admin           |                                         | {}
  can_create_db   | CREATEDB, NOLOGIN                       | {}
  can_create_role | CREATELOGIN, CREATEROLE, NOLOGIN        | {}
  root            |                                         | {admin}
(4 rows)
```

You have the following options for roles:

- CREATEROLE: Allows the role to CREATE, ALTER, and DROP non-admin roles
- LOGIN: Allows the role to log in using client authentication
- CONTROLJOB: Allows the role to pause, resume, and cancel jobs
- CONTROLCHANGEFEED: Allows the role to create CHANGEFEED on tables where it has the SELECT privilege
- CREATEDB: Allows the role to create or rename a database
- VIEWACTIVITY: Allows the role to execute SHOW STATEMENTS and SHOW SESSIONS, including queries that are executed by other roles
- CANCELQUERY: Allows the role to cancel a query
- MODIFYCLUSTERSETTING: Allows the role to modify cluster settings
- PASSWORD <password>: Allows a role to use a given password to authenticate itself
- VALID UNITL <timestamp>: Specifies the validity of a given password

You can prepend the NO keyword to many of these roles to take away a given role. For example, NOCREATEROLE will not allow a role to create or manage non-admin roles. Similarly, NOCREATEDB will not allow a role to create or rename a database.

Privileges

You can use the GRANT command to give specific privileges to a given role. In the following example, we are providing complete privileges to the admin role on the startrek database:

```
$ GRANT ALL ON DATABASE startrek TO admin;
```

Now, you can look at all the grants on the startrek database and see that the admin role has ALL access:

```
$ SHOW GRANTS ON DATABASE startrek;
  database_name | grantee | privilege_type
----------------+---------+----------------
  startrek      | admin   | ALL
  startrek      | root    | ALL
```

There are various levels where grants can be given, such as the database, schema, table, and type level. The `ALL` keyword specifies all the levels in the schema hierarchy.

Privilege can be given on the following constructs:

- `CREATE`
- `DROP`
- `SELECT`
- `INSERT`
- `DELETE`
- `UPDATE`
- `USAGE`
- `ZONECONFIG`
- `CONNECT`

You can also manage users in the admin console in the cluster's **SQL Users** page.

In the next section, we will discuss data encryption at rest and in flight and how to configure the same.

Data encryption at rest and in flight

Encryption is the process of encoding plain text into an alternative unreadable format known as ciphertext. Decryption is the process of decoding the ciphertext back into its original plain text readable format. It is important to encrypt stored data, as well as the data that's being transferred between the client and nodes. In this section, we will learn how to achieve this.

Encryption at rest

Data at rest indicates the data that is stored on a physical storage system, such as a disk. **Encryption at rest** is an Enterprise-only feature. This feature allows you to encrypt all the files on the physical storage using **Advanced Encryption Standard** (AES).

Two types of keys are involved:

- **Store keys**: These are provided by the user and are used to encrypt data keys.
- **Data keys**: These are generated by CockroachDB and are used to encrypt all the files on disk. They are persisted in a registry file and are encrypted using the store key.

You can generate an AES-128 encryption key using the following command:

```
$ cockroach gen encryption-key -s 128 encr-keys/encryption.key
successfully created AES-128 key: encr-keys/encryption.key
```

Now, you can use the encryption key that you generated in the previous step in the input when you are starting the node:

```
$ cockroach start \
--insecure \
--store=node4,attrs=encrypt \
--enterprise-encryption=path=node4,key=encr-keys/encryption.
key,old-key=plain \
--listen-addr=localhost:26259 \
--http-addr=localhost:8082 \
--join=localhost:26257,localhost:26258,localhost:26259 \
--locality=region=us-west
--background
```

Here, when the node is started, it uses the old key to read the data and then rewrites the data using the new key. If you want to disable the encryption, you can specify key=plain as the encryption configuration.

Now, let's look at encryption in flight.

Encryption in flight

CockroachDB uses TLS 1.2 to encrypt the client-node and inter-node communication. Please refer to the *Client and node authentication* section to learn how to configure certificates and keys for the TLS. If you don't need encryption for client-node communication, you can use the --accept-sql-without-tls flag when you are starting the node. This lets the node accept connections from clients without TLS.

In the next section, we will learn how to enable SQL audit logging to monitor activities on a table.

Audit logging

SQL audit logging is an important security feature that you can use to track all the activities that are occurring in a given CockroachDB cluster. Specifically, you can select tables whose activity must be tracked and only enable audit logging on them.

The following information gets logged during auditing:

- Full query text.

- The date and time of the query.

- The client's IP address.

- The application's name.

- The user.

- The event type, which will be SENSITIVE_TABLE_ACCESS. This indicates that it's an event related to SQL audit logging.

- The name of the table that was queried.

Now, let's look at an example. We will enable audit logging for one of the tables in the default databases. startrek is a database that comes by default with the open source CockroachDB:

```
$ show databases;
  database_name | owner | primary_region | regions | survival_
goal
----------------+-------+----------------+---------+----------
------
  defaultdb     | root  | NULL           | {}      | NULL
  postgres      | root  | NULL           | {}      | NULL
  startrek      | root  | NULL           | {}      | NULL
  system        | node  | NULL           | {}      | NULL
```

Let's enable SQL audit logging for the episodes table, as shown here:

```
$ ALTER TABLE startrek.episodes EXPERIMENTAL_AUDIT
SET READ WRITE;
ALTER TABLE
```

Now, if you look at one of the latest entries in the CockroachDB SQL audit log, you should see an entry that indicates that SQL audit logging has been enabled for the `startrek.` `episodes` table. Usually, this SQL audit log can be found under the logs directory and the log file will be `cockroach-sql-audit.log`. The following is an entry from the SQL audit log that shows the `ALTER TABLE` statement that enabled SQL audit logging:

```
I211129 01:03:01.155934 25421 8@util/log/event_log.go:32
: [n1,client=<127.0.0.1:58274>,hostnossl,user=root] 3
={"Timestamp":1638147781150167000,"EventType":"sensitive_
table_access","Statement":"<ALTER TABLE startrek.public.
episodes EXPERIMENTAL_AUDIT SET READ WRITE>","Tag":"ALTER
TABLE","User":"root","DescriptorID":53,"ApplicationName":"$
cockroach sql","ExecMode":"exec","Age":5.706,"TxnCounter":
16,"TableName":"<startrek.public.episodes>","AccessMode":"rw"}
```

Now, let's execute a `SELECT` statement against the same table:

```
$ select * from startrek.episodes;
```

Now, if you go back to the `logs/cockroach-sql-audit.log` file once more and look at the latest entries, you should see an entry for the `SELECT` statement that you executed:

```
I211129 01:03:38.010291 25421 8@util/log/event_log.go:32
: [n1,client=<127.0.0.1:58274>,hostnossl,user=root] 4
={"Timestamp":1638147818007255000,"EventType":"sensitive_
table_access","Statement":"<SELECT * FROM \"\".startrek.
episodes>","Tag":"SELECT","User":"root","DescriptorID":53,
"ApplicationName":"$ cockroach
sql","ExecMode":"exec","NumRows":79,"Age":2.01,"FullTableScan
":true,"TxnCounter":18,"TableName":"<startrek.public.
episodes>","AccessMode":"r"}
```

If you no longer need SQL audit logging for a particular table, it's possible to turn it off, like so:

```
ALTER TABLE startrek.episodes EXPERIMENTAL_AUDIT SET OFF;
```

You can confirm that audit logging has been turned off by going through the entries in the SQL audit log file; that is, `logs/cockroach-sql-audit.log`:

```
I211129 01:15:32.779808 25421 8@util/log/event_log.go:32
⋮ [n1,client=<127.0.0.1:58274>,hostnossl,user=root] 9
={"Timestamp":1638148532758348000,"EventType":"sensitive_
table_access","Statement":"<ALTER TABLE startrek.public.
episodes EXPERIMENTAL_AUDIT SET OFF>","Tag":"ALTER
TABLE","User":"root","DescriptorID":53,"ApplicationName":"$
cockroach sql","ExecMode":"exec","Age":12.184,"TxnCounter":30,
"TableName":"<startrek.public.episodes>","AccessMode":"rw"}
```

In the next section, we will learn about RTO and RPO, two of the key parameters in defining your stance for data protection, data loss, and business disruption.

RTO and RPO

RTO determines how soon you can recover from a disaster and start serving requests. It's almost impossible to have zero RTO, which means there will be some amount of application downtime whenever things go wrong. RPO determines how much data you can lose during a failure without causing any major business impact.

Technically, having a very low RTO and zero RPO is the dream of any team that manages the database, but it's incredibly hard to achieve. Also, having no data loss is an important requirement for many mission-critical applications that can never lose any committed data. Since we are talking about data at scale, including several nodes where data is being replicated multiple times, nodes that are sitting in different cloud regions, and infrastructure spread across heterogeneous systems, it's very challenging to achieve desirable RTO and RPO numbers.

There is an Enterprise-only feature that you can use to take full and incremental backups. Incremental backups determine your RPO number as you can always restore to the previous incremental backup. Also, if your data becomes corrupted, you can use a **point-in-time restore** to restore your cluster to a specific timestamp. You can also pay extra attention to the storage layer and make it highly redundant and highly available so that there is absolutely no data loss at the storage layer. You can achieve zero RPO with a full and incremental backup strategy and having enough replicas across availability zones and regions.

Whenever we have a non-zero RTO, this means that there is some amount of downtime. Depending on the nature of the business, we can significantly lose income during such disruptions. It also affects the overall reputation of the company. Luckily, with the cloud, it's relatively easier to provide redundancy in terms of compute, network, memory, and storage. If everything is automated correctly, we should achieve a few seconds of RTO.

In the next section, we will discuss network security.

Keeping the network secure

Networks are the most popular places for hackers to perform targeted attacks. With today's modern infrastructure being comprised of on-premises, private, public cloud, and multi-cloud environments, there are ample opportunities for hackers to get into insecure networks. First, we should ensure that all the communication that goes in and out of the CockroachDB cluster is completely secured and encrypted. It's always a good idea to turn on TLS for inter-node and client-node communication. Once the data becomes larger, we will end up having a dedicated **Site-Reliability Engineering** (**SRE**) organization that ensures CockroachDB is up and running at all times. We should ensure that the right set of folks has the right access to the data. DDLs such as DROP and ALTER should be much more restrictive in production. Also, at any given time, only the folks on production on-call rotation should have access to bastion hosts.

Wherever we have deployed the CockroachDB instances, whether it's directly on a virtual machine, a bare-metal server, or a Docker container, we have to make sure that only the relevant ports can be accessed and that only secured connections are allowed. Also, if you are using a service mesh such as Istio, it already provides out-of-the-box mTLS and TLS termination proxy services.

It's important to only authorize specific application networks that can access the CockroachDB cluster. This can be achieved through **IP allowlisting** and **virtual private cloud** (**VPC**) peering. In IP allowlisting, you provide the IP addresses of all the applications that need to access the CockroachDB cluster.

In VPC peering, you connect two virtual private clouds so that all the traffic between them can be routed using private IP addresses. The advantages of VPC peering are as follows:

- Improved network latency as the traffic between two virtual private clouds doesn't have to go through the public internet

- More secure as the traffic is isolated from the public internet

- More cost savings as we will avoid using external IP addresses, which reduces the egress cost

In the next section, we will look at some of the security best practices that should be followed to get maximum security.

Security best practices

Let's go over some of the best security practices:

- **Certification and key rotation**: It is important to rotate certificates and keys from time to time and keep their expiration times shorter. We should also automate a way to rotate certificates and keys, which will be useful whenever we come across any attacks. There are secret management tools such as Vault that can make it easy to automate dynamic certificate generation and rotation.

- **Client password**: We should ensure that we follow all the necessary rules to generate a very strong password. Weak passwords are easier to predict, which makes them more vulnerable.

- **Planning for disaster recovery**: We should be diligent about our backup and restore strategy, ensuring that none of the backups are missed and that they are stored in multiple different regions.

- **Automation**: It's always a good practice to automate most of the routine work around security. This will allow us to quickly respond without too much manual intervention when things go wrong.

- **Data encryption**: It's important to encrypt data at rest as hackers constantly try to get access to the storage devices.

- **Transport layer security** (**TLS**): Unless both the client and cluster nodes are behind a firewall or within a secure perimeter, it's always a good idea to enable TLS.

- **Secret management**: Several tools are available for managing secrets. So, please make use of any of the well trusted solutions, rather than reinventing the wheel.

- **Production access**: Production access should be highly restricted and should be done on an as-needed basis. We must make sure that only the folks who are currently on production on-call rotation have access, not everyone. Setting up bastion hosts to access SQL clients is a must.

- **Backups and archived data**: We must make sure that even the backups and historically archived data are encrypted.

- **Personally identifiable information** (**PII**): We should make sure that the PII and sensitive information doesn't show up in some database logs or audit logs.

With that, we have come to the end of this chapter.

Summary

In this chapter, we looked at several aspects of security, including authentication, authorization, encryption, disaster recovery, auditing, and network security. Since a given CockroachDB infrastructure can be spread across on-premises environments and various private and public cloud providers, it's important to ensure maximum security and be prepared to quickly recover in case things go wrong. Hackers and ransomware attacks are increasing every day, so being educated about security and constantly improving our security posture is the only way to prevent attacks.

In the next chapter, we will discuss debugging various performance-related issues in CockroachDB.

10
Troubleshooting Issues

In *Chapter 9, An Overview Of Security Aspects*, we learned about various mechanisms for securing SQL workloads on CockroachDB. The ever-increasing presence of software running on the cloud has led to increased attack surfaces. So, it is important to carefully examine your security posture and to take measures to fill any security gaps.

In this chapter, we will go over how to troubleshoot issues that you come across while using CockroachDB. Since this is a very broad topic, we will cover a few examples from each category. Debug logs can help in narrowing down possible trouble areas and eventually finding the root cause. In the first section, we will learn about collecting debug logs. Next, we will go through some of the common causes of connection issues. In the later sections, we will cover various topics that can cause a query to execute slowly or fail completely. We will also go over ideal resource allocation and general guidelines to be followed during an upgrade and advanced debugging options.

The following topics will be covered in this chapter:

- Collecting debug logs
- Connection issues
- Tracking slow queries
- Capacity planning

- Configuration issues
- Network latency
- Guidelines to avoid issues during upgrade
- Advanced debugging options

Technical requirements

There are some commands discussed in this chapter that require you to have CockroachDB installed. If you still haven't done so, please refer to the *Technical requirements* section in *Chapter 2, How Does CockroachDB Work Internally?*

We will first start with a discussion on collecting debug logs.

Collecting debug logs

Whenever we debug issues, it's very helpful to have aggregated logs. There are tools such as **Datadog** that provide a single pane of glass for monitoring a CockroachDB cluster. Node logs and the admin UI provide helpful information for troubleshooting issues. Please refer to *Chapter 8, Exploring the Admin User Interface*, to get yourself familiar with the user interface. Logs provide detailed information about all the activities happening in a cluster. CockroachDB provides various types of log files based on the intent. CockroachDB also supports log levels and log channels. It is also possible to emit certain log messages to an external destination for further processing.

In this section, we will first discuss various log files and what information each file contains, then move on to understanding the log levels that determine how much information you want to log based on their severity. Later, we will learn about the concept of log channels, sending logs to an external resource for further processing, and collecting all the logs.

Log files

CockroachDB provides various node-level log files that specialized in certain things, such as the health of nodes, storage engine logs, or security-related logs. Following is the list of various log files that are available in CockroachDB:

- `cockroach.log`: A general log that contains information about all the major events occurring within the cluster.
- `cockroach-health.log`: The health of the cluster.

- `cockroach-security.log`: SQL security log.

- `cockroach-sql-audit.log`: SQL access audit log.

- `cockroach-sql-auth.log`: SQL authentication log.

- `cockroach-sql-exec.log`: SQL execution log.

- `cockroach-sql-slow.log`: SQL slow query log.

- `cockroach-sql-schema.log`: SQL schema change log.

- `cockroach-pebble.log`: Pebble key-value store log. Pebble is the default storage engine in CockroachDB.

- `cockroach-telemetry.log`: Telemetry log, which contains event information about feature usage.

Next, we will discuss log levels.

Log levels

Log levels indicate the severity of a log message and based on the severity you have to decide whether they should be handled or not. In general, the log levels `ERROR` and `FATAL` should be handled in your application logic and it should take appropriate action based on the error type. Following are the four log levels in CockroachDB:

- `INFO`: This indicates general informational log messages. It might be prudent to turn these off in a production environment as they can occupy a lot of space and are not that relevant when debugging issues.

- `WARNING`: This indicates that a normal operation might have failed but will recover automatically. Based on each case and the impact, we have to decide whether to handle this or not.

- `ERROR`: This indicates that a normal operation cannot be performed. We should pay attention to these errors and make sure they are handled properly.

- `FATAL`: Fatal errors need immediate attention and action as they will be of the highest severity.

Next, we will learn about a concept called log channels.

Log channels

Log channels are distinguished based on the type of operation rather than the severity. A log channel is useful when you have multiple teams that have to look at different sets of logs based on the team's responsibilities. For example, logs related to configuration changes done by an admin user in production will be more useful to an infosec or security team than an application developer team. Log channels are very useful when integrated with a sink such as Slack. A sink is an external resource that is capable of receiving data. For example, all OPS and HEALTH messages can go to a **Site Reliability Engineering (SRE)** channel, whereas USER_ADMIN, PRIVILEGES, and SENSITIVE_ACCESS can go to the infosec channel. Following is a list of available log channels in CockroachDB:

- DEV: This channel is used during development and everything gets logged.
- OPS: This channel is related to cluster relation operations, configurations, and jobs.
- HEALTH: This channel logs resource usage, connection errors, range availability, and leasing events.
- STORAGE: This is used to log Pebble storage engine related events.
- SESSIONS: This covers sessions, connections, and authentication events.
- SQL_SCHEMA: This is used to track schema changes involving database, schema, table, sequence, view, and type; metadata changes.
- USER_ADMIN: This is used to track changes in users and roles.
- PRIVILEGES: This is used to track changes in grants and object ownership.
- SENSITIVE_ACCESS: This is used to access data access audit events, SQL statements by admin, and operations that write to system tables.
- SQL_EXEC: This channel logs SQL executions and errors during execution.
- SQL_PERF: This channel records events that affect performance and slow query logs.
- SQL_INTERNAL_PERF: This logs internal details of CockroachDB that will be useful during troubleshooting.
- TELEMETRY: This is used to log telemetry events.

Next, we will discuss sending logs to an external sink.

Emitting logs to an external sink

A **sink** is an external resource that is capable of receiving data. This external resource sits outside of CockroachDB and is deployed and maintained separately. It is possible to route messages of one or more log channels to an external log sink, which can be used for alerting, providing an aggregated view of the logs across all the nodes in the cluster, and further analyzing the log data, which can be useful in automating responses to unexpected or fatal events. Sinks include log files, Fluentd compatible servers, HTTP servers, and standard error streams. With appropriate configuration, it is possible to redact sensitive data while sending it to the sink. Sinks support the following parameters:

- `filter`: The minimum severity level.

- `format`: The log format.

- `redact`: If true, redacts sensitive data such as **personally identifiable information** (**PII**) from log messages.

- `redactable`: Retains redaction markers around sensitive fields.

- `exit-on-error`: If true, stops the CockroachDB node if it's unable to send log messages to the sink.

- `auditable`: If true, enables exit-on-error and disables buffered-writing and hence enforces stopping the CockroachDB node.

- If it's unable to send logs to the sink, it flushes each entry and synchronizes the writes.

Additionally, it is advisable to go through technical advisories that report major issues in CockroachDB related to security and correctness. These can be accessed at `https://www.cockroachlabs.com/docs/advisories/index.html`. Also, Cockroach Labs keeps updating newly seen issues and resolution details at `https://www.cockroachlabs.com/docs/stable/cluster-setup-troubleshooting.html#`.

Gathering Cockroach debug logs

`cockroach debug zip` connects to the cluster and collects the debug information from all the nodes in the cluster. So, when you start troubleshooting, it would be helpful to run this to collect all the relevant logs. Following is the syntax of the `debug` command:

```
cockroach debug zip <path_to_store_the_zip_file> { flags }
```

The following is an example of the `debug` command:

```
$ cockroach debug zip ./cockroach-db/logs/debug.zip --insecure
--host=localhost:26258
```

Following are some of the important options that you have with the `cockroach debug zip` command:

- `--redact-logs`: Redacts sensitive information from the logs.
- `--timeout`: Times out the command with an error message if it doesn't complete within the stipulated time.
- `--files-from` and `--files-until`: You can filter the logs based on the time. For example, `--files-from='2021-01-01 12:00'` and `files-until='2021-06-30 12:00'`.

For complete options, please check `cockroach debug zip --help`.

Information such as CPU usage, metrics, schema metadata, node health, stack trace, events, and jobs are collected from all the nodes.

Next, we will learn about connection issues and how to ensure we don't run into one.

Connection issues

Connection refused is one of the common connection issues when a PostgreSQL-compatible SQL client is unable to connect to a CockroachDB node. There can be many things that can go wrong here, such as expired certificates or firewall settings. Following are some general guidelines on what to look for in terms of connection related configurations:

- Check that the CockroachDB node is running and listening on the correct port.
- Check the connection details, especially the hostname and port number.
- If the node is running in secure mode, you have to make sure appropriate certificates and keys are generated and are being passed correctly during the connection.
- Make sure the client certificate is present and not expired.
- If the node is already running and the host and port are correct, you can try restarting the node and see if that helps with the connection issue.
- Check for firewall rules that prohibit specific inbound and outbound traffic.
- Make sure the node is reachable using `ping` and the port can be accessed using debugging tools such as *netcat* or *telnet*.

In case of connection issues, we have to make sure that the client and the nodes of a cluster can talk to each other securely.

In the next section, we will go over some of the recommended hardware and software configurations for CockroachDB nodes.

Tracking slow queries

It is important to know which queries are not performing as expected so that we can further investigate to find the actual cause for the slowness. Some queries can become slow when the data grows. So, it is important to benchmark some of the critical queries over time to make sure they are getting executed within the expected time. There are three major ways to identify and debug slow queries:

- Use EXPLAIN and see if the query involves full scans and if so, see what kind of indexes you can create to avoid it. This holds true for table joins as well.

- Turn on the slow query log for a specific latency threshold. In the following example, all queries whose latency is greater than or equal to 500 milliseconds will be logged in the slow query log:

```
> SET CLUSTER SETTING sql.log.slow_query.latency_
threshold = '500ms';
SET CLUSTER SETTING
Time: 120ms total (execution 120ms / network 0ms)
```

- Integrate CockroachDB logging with tools such as OpenTelemetry and Jaeger that provide advanced tracing and spanning capabilities. This helps us with understanding all the network hops that happen during a query and also how much time individual units of work are taking. Whichever function or network hop takes more time could be a potential candidate to investigate further.

A slow query can be caused by various issues that are discussed in the rest of the sections in this chapter. So, once you identify a slow query, you have to identify which part of the query is slow and investigate further till you find all possible reasons.

Next, we are going to go over general guidelines for upgrading a CockroachDB cluster.

Capacity planning

We have to make sure individual nodes and the entire cluster has optimal hardware and software configuration in order to perform better. Following is a set of recommended configurations for CockroachDB:

- **Operating system**: Container-optimized OSs such as Ubuntu, Red Hat Linux, and CentOS are the preferred operating systems. A container-optimized OS, as the name suggests, is an operating system that is optimized for running Docker containers.

- **Node configuration**: Following are node configurations for optimal performance:

  - **CPU**: At least four vCPUs per node. For more throughput per node, you can increase the number of vCPUs.

  - **Memory**: 4 GiB RAM per vCPU.

  - **IOPS** (input/output operations per second): 500 per vCPU.

  - **Disk I/O in MiB/second**: 30 MiB/second per vCPU.

- **Storage**

  - 150 GiB per vCPU.

  - 2.5 TiB max per node.

  - Dedicated volumes only used for CockroachDB are preferred.

VM capacities differ between cloud providers. So, based on the preceding recommendations, you can choose the appropriate VM for your workloads.

It's important to monitor the CPU and storage metrics for all the nodes and to have some automation in place in order to expand the cluster by adding more nodes. Although there is no hard limit, you might be better off keeping the maximum number of nodes in a cluster to 50. Once you hit this threshold, you can add a second separate cluster and introduce an external sharding mechanism to decide which cluster to redirect to for a given query.

In the next section, we will go through configuration-related issues.

Configuration issues

Things can go wrong because of misconfigurations such as using unavailable ports and configuring the incorrect storage directory at the start. CockroachDB provides several cluster-level settings. In most cases, leaving the default values might work better. If you are modifying any default value, please ensure it's tested at scale in a pre-production environment. Following are some of the commonly seen configuration-related issues:

- **Storage directory already exists**: This happens when you try to start a node with a storage directory that has been used for some other CockroachDB process. In this case, you can either choose a different directory or delete the contents of that current directory.

- **Port is already in use**: This happens when there is some other process that is already using a given port with which you are trying to start a CockroachDB process. You can either kill the process that is using that port, provided that process is no longer required, or you can also pick a different port that is available.

- **Clock synchronization error**: Whenever the clock of some node goes out of sync with at least half of the other nodes in the cluster by more than 500 milliseconds (the default threshold), the node shuts itself down. In order to avoid this, you can try using an external **network time protocol** (**NTP**) service.

- **Open file descriptor limit**: Since CockroachDB opens a large number of file descriptors, it expects a node to have a certain threshold limit, which is 1,956. The recommendation is to set the limit to *unlimited*.

There are many other issues that you can face related to configuration. Based on the error message, you have to further investigate and decide the appropriate solution.

In the next section, we will discuss slow queries.

Guidelines to avoid issues during an upgrade

When upgrading your cluster, the following are some general guidelines:

- Before you upgrade to a new version, it would be good to wait at least 2–3 months to make sure it's stable enough, doesn't contain too many bugs, and is not withdrawn for security or functional issues.

- Once you decide the version you want to upgrade to, you have to make sure you can upgrade to that version from the current version after going through the release notes. Sometimes you might have to upgrade multiple times to reach the desired version.

- For certain versions, it's not possible to downgrade back to the previous version if things go wrong and the upgrade is auto-finalized. In that case, your only option is to discard the current cluster and create a new cluster from the backups. So make sure all the data is backed up before you start the upgrade. Also, as an alternate solution, you can disable auto-finalization with the following command:

```
SET CLUSTER SETTING cluster.preserve_downgrade_option =
'<current_cluster_version>';
```

However, you have to make sure you manually re-enable auto-finalization after ensuring the cluster is stable and there is no data corruption. The following command can be used for re-enabling auto-finalization:

```
RESET CLUSTER SETTING cluster.preserve_downgrade_option;
```

- You can perform a rolling upgrade by upgrading one node at a time and letting the upgraded node rejoin the original cluster. Before you upgrade the next node, make sure the cluster is healthy.

- Make sure you are not decommissioning any nodes during the upgrade. This can result in multiple failures.

- It is also advisable not to do any schema changes.

- After the upgrade is complete, you can execute the `Show Cluster Setting` version to ensure it shows the latest upgraded version.

- If anything goes wrong during the upgrade, your best option is to run the `'cockroach debug zip'` command in order to gather all the information about the cluster and go through the failures.

In the next section, we will discuss network latency.

Network latency

In the admin user interface, you have a page that shows network latencies within the cluster. It has information about the round-trip latencies between each pair of nodes in the cluster as shown in the following screenshot.

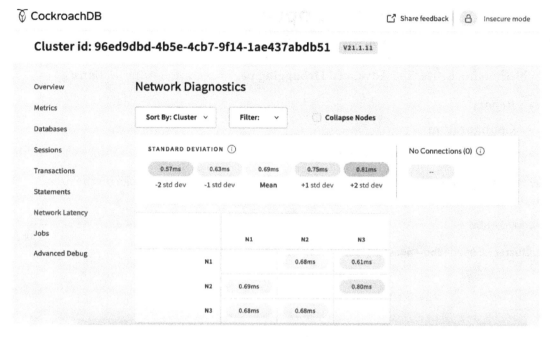

Figure 10.1 – Network latency page in the admin user interface

Network latency plays an important role when you are deciding on your replication strategy and topology patterns. Please refer to *Chapter 4, Geo-Partitioning,* to go over all the topology patterns available in CockroachDB. Based on the criticality of the data, you also have to decide whether you can tolerate zonal or regional failures.

Also, even before you decide the region and availability zone for CockroachDB nodes, it would be good to go through inter-region network latencies for different cloud providers. If you are planning to have your CockroachDB cluster span across multiple cloud providers, you may want to choose the same region for your second and third cloud provider in order to ensure the least network latency as they will be geographically co-located. Companies such as Aviatrix also provide inter-cloud latencies for various regions.

In the last section, we will go over some of the advanced debugging options.

Advanced debugging options

In the admin user interface, you have a page for advanced debugging, which, as the name indicates, shows advanced information about a cluster that can be useful in troubleshooting issues. The **Advanced Debugging** page has the following sections:

- **Reports**
- **Configuration**
- **Even more advanced debugging**

The following screenshot shows how the **Advanced Debugging** page looks on the user interface:

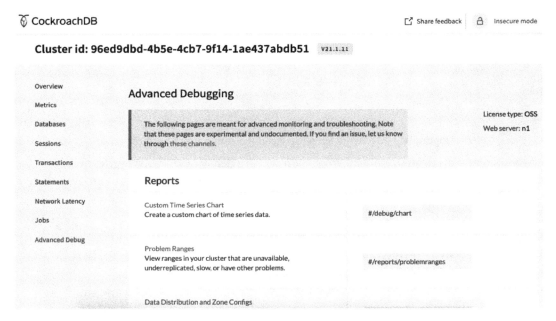

Figure 10.2 – Advanced Debugging page in the admin user interface

Under **Reports** you can get a report for the following items:

- **Custom Time Series Chart**: You can create a custom chart of the time series data.
- **Problem Ranges**: You can view ranges in your cluster that are unavailable, under-replicated, slow, or have other problems.

- **Data Distribution and Zone Configs**: You can view the distribution of table data across nodes and verify the zone configuration.

- **Statement Diagnostics History**: You can view the history of statement diagnostics requests.

The **Configuration** section shows the following items:

- **Cluster Settings**: You can check all the cluster-level configurations here.

- **Localities**: This shows node locality, including address and location.

The **Even More Advanced Debugging** section is meant for the developers of CockroachDB who understand the internal implementation. It has a lot of information, is too detailed, and in general, can be avoided while debugging issues unless you understand the internal implementations of CockroachDB.

Summary

In this chapter, we went through several resources that are available for debugging and troubleshooting issues. Getting yourself familiarized with the debug logs and admin user interface is an important step if you want to quickly diagnose and root cause the issues. Also, being aware of cluster configurations, deployment strategies, network latencies, and security settings comes in handy while troubleshooting issues. Using tools for log aggregation, distributing tracing, tracking spans, alerting, and advanced metrics can further aid you.

In the next chapter, we will learn about performance benchmarking and migration from other databases to CockroachDB.

11
Performance Benchmarking and Migration

In *Chapter 10, Troubleshooting Issues*, we learned how to troubleshoot issues with the help of metrics and logs. In this chapter, we will go over some important parameters that we should consider during performance benchmarking, how to do performance benchmarking, and important things to consider during migration. Finally, we will look at some specific examples of migrating from some traditional databases, such as **PostgreSQL**.

The following topics will be covered in this chapter:

- Performance—Things to consider
- Performance benchmarking for CockroachDB
- Migration—Things to consider
- Migrating from traditional databases

Technical requirements

There are some commands discussed in this chapter that require you to have CockroachDB installed. If you still haven't done so, please refer to the *Technical requirements* section of *Chapter 2, How Does CockroachDB Work Internally?*.

Performance – Things to consider

In this section, we will discuss general factors related to the infrastructure that affect performance, some standard benchmarking suites, and running a benchmark for your specific needs in CockroachDB.

Infrastructure

It is important to know the ideal configuration for your infrastructure as that significantly influences the performance of CockroachDB that runs on top of it. Here are some key factors to consider in this regard:

- **Central processing unit (CPU)**: It would be a cliché to say that the CPU plays an important role in performance. We have to ensure each node gets at least four **virtual CPUs (vCPUs)**. You should constantly monitor the CPU usage in all nodes to quickly identify hot nodes that might be receiving more traffic than other nodes. This can make the node's CPU reach its maximum capacity, which can slow down query response times.

- **Memory**: Pebble key-value store, table metadata, and CockroachDB internal data structures all use memory. It is recommended that each node should have at least 4 **gibibytes (GiB)** of **random-access memory (RAM)**. Also, you should keep track of the memory usage of all nodes. If a node or bunch of nodes start to reach their full memory capacity, you can either increase the memory of the affected node/s or scale out the cluster by adding more nodes. **Resharding** the ranges ensures that each node now handles fewer ranges, and hence their memory usage can go down. Also, ideally, 20% of the memory should always be empty, so you should treat reaching 80% memory usage as the full capacity and start readjusting the cluster if any nodes reach that threshold.

- **Operating system**: Any Linux-based operating system would be ideal. Container-optimized operating systems, Ubuntu, **Red Hat Enterprise Linux (RHEL)**, and CentOS are among the popular choices for an operating system for running CockroachDB.

- **Storage input/output (I/O)**: Many cloud providers offer storage options that are highly available. You can also choose specific configurations to ensure the storage layer withstands zonal and regional failures. **Solid-state drives (SSDs)** are preferred for faster I/O. You should not exceed 2.5 tebibytes (TiB) per node and 150 gigabytes (GiB) per vCPU for optimal performance, as per the recommendation from Cockroach Labs..

- **Network**: Since more than one node is involved in a distributed **Structured Query Language (SQL)** query and SQL clients can also consume data from various geographies, network performance is critical in determining the overall performance. If a cluster is spread across multiple regions and supports regional failures, then the overall query-response time can be slower, as each commit would have to be communicated between at least two regions. So, pay attention to network hops between the nodes, client, and cluster.

- **Cloud provider**: We now have several cloud providers that offer compute, memory, storage, and network resources on the cloud. Also, you get several options to choose from for each of these resources. There is also an increasing use of hybrid cloud and multi-cloud architectures, which further complicates the deployments. Cloud providers also refresh their hardware; so, when you are benchmarking across multiple cloud providers, you might see inconsistencies in performance if you repeat the same benchmarks every 6 months.

Next, we will learn about some popular benchmark suites.

Popular benchmark suites

Several popular benchmark suites are used to benchmark databases and data warehouses. Here is a list of some commonly used ones:

- **TPC-C** (http://www.tpc.org/tpcc/): TPC-C is a short form of **Transaction Processing Performance Council Benchmark C**, which is a benchmark used to compare the performance of **online transaction processing (OLTP)** systems. The TPC-C benchmark defines a set of functional requirements that can be run on a given transaction processing system. This benchmark requires reporting performance, which is the same as transaction throughput, and price/performance, which is the overall infrastructure cost/throughput.

- **TPC-H** (http://www.tpc.org/tpch/): TPC-H is a decision support benchmark. It involves running ad hoc queries and concurrent data modification and measuring performance.

- **YCSB** (`https://github.com/brianfrankcooper/YCSB`): Short for **Yahoo! Cloud Serving Benchmark**, YCSB is an open source benchmark suite used for evaluating retrieval and maintenance capabilities of databases. It is often used to measure the performance of NoSQL databases and workloads involving simple database operations.

- **Sysbench** (`https://github.com/akopytov/sysbench`): Sysbench is a benchmark suite used for measuring system performance, especially when you are running a database.

It takes a lot of effort and resources to run these benchmarks. Fortunately, Cockroach Labs conducts extensive benchmarking across multiple cloud providers every year that includes the performance of CPU, network, storage, and TPC-C, and provides the results in its annual cloud report. Benchmarks for 2020 and 2021 can be accessed at `https://www.cockroachlabs.com/guides/2020-cloud-report/` and `https://www.cockroachlabs.com/guides/2021-cloud-report/`. So, unless you want to measure the performance for some specific use cases and workload patterns, these cloud reports should serve as a great starting point to decide which cloud provider to pick and if CockroachDB's performance is acceptable.

Benchmarking your specific use cases

In any benchmark, you should also pay attention to whether you want to conduct a synthetic benchmark or a real-world benchmark. **Synthetic benchmarking** involves simulating the traffic with dummy data. **Real-world benchmarking** is much closer to what you already have in production, benchmarking the real workloads with the real data. Real-world benchmarks can give a much better indication of how CockroachDB is going to perform when put in production. Although there are standard suites available for performance benchmarks, it's always handy to do your own benchmarking for specific needs. Here are some categories that can fall under this:

- **Large data**: When the data is large, you would need a greater number of nodes to manage the data—so, the performance of a 3-node cluster might vary from the performance of a 17-node cluster. So, based on the specific volume of data that you have in your mind, you can benchmark CockroachDB against it. You should also ensure that the read-and-write queries are similar to what you would expect in production, in order to keep it close to the real world.

- **High traffic**: Your workload might have to be more skewed toward high read or high write traffic, or both. So, based on what is relevant to your query patterns, it's important to benchmark for high traffic workloads tailored to your needs.

- **Scaling with an increase in data volume**: If you are using a database in a year-old start-up with very few customers, the volume of data will obviously be much less. But as the company grows and there are more use cases and customers start consuming more data, the data volume can explode. So, it is good to prepare your infrastructure for 5-10X growth, both in terms of data volume and traffic, and see how many resources you would need to scale out. As part of this exercise, you will also know where you stand with respect to automation. By automating all the repeatable tasks, you can be better prepared to handle failures and scale out the infrastructure.

- **Failover**: Some businesses are more sensitive to downtime than others. For example, if your database is supporting some application used by doctors and health specialists to view patient records in real time or if your database supports an application that is used for online booking, your database is expected to be available maybe with six nines of availability. This means the database can be unavailable only for 31.56 seconds in the entire year. So, in such cases, it's important to simulate failures to see how quickly you can recover from the node, zonal, regional failures, or any kind of failures that can result in the disruption of your database service.

- **Multi-tenancy**: There are a couple of different ways to implement multi-tenancy with CockroachDB. For example, you can have a dedicated small cluster for each customer. Although this is great from a security perspective as workloads from other customers can be completely isolated, practically, this is not cost-effective. So, it would be ideal to just introduce `tenant-id` (**tenant identifier** (**ID**)) and shard the data based on that. But with this approach, it can become tricky to keep the **service-level agreement** (**SLA**) the same for all customers, as it's more prone to creating hotspots as the tenants will not be equally active in consuming the data. In such cases, you can also benchmark resharding ranges or try resharding the data into multiple clusters, and see how long that takes.

- **Migration**: Before you migrate the real data in production, you can benchmark how much time the entire process takes from exporting the schema and data from the source database to importing the same into the target database. You might have to do this for various data volumes to get an idea of how much time it might take in production for the complete migration.

In the next section, using an example, we will learn how to conduct performance benchmarking for CockroachDB.

Performance benchmarking for CockroachDB

In this section, we will go over the process of performing TPC-C benchmarking on CockroachDB. Here are the steps for running a TPC-C workload and getting the statistics for the run:

1. Import the TPC-C dataset into your local CockroachDB cluster, as follows:

```
$ cockroach workload fixtures import tpcc —warehouses=5
'postgresql://root@localhost:26257?sslmode=disable'
I220127 06:46:52.189260 1 ccl/workloadccl/fixture.go:342
[-] 1  starting import of 9 tables
```

2. After loading the tables, you should see the following statistics at the end:

```
I220127 06:47:31.384005 1 workload/tpcc/tpcc.go:485    [-]
13   check 3.3.2.1 took 152.063ms
I220127 06:47:31.664548 1 workload/tpcc/tpcc.go:485    [-]
14   check 3.3.2.2 took 280.468ms
I220127 06:47:31.719888 1 workload/tpcc/tpcc.go:485    [-]
15   check 3.3.2.3 took 55.281ms
I220127 06:47:32.796841 1 workload/tpcc/tpcc.go:485    [-]
16   check 3.3.2.4 took 1.076892s
I220127 06:47:33.218333 1 workload/tpcc/tpcc.go:485    [-]
17   check 3.3.2.5 took 421.432ms
I220127 06:47:34.706096 1 workload/tpcc/tpcc.go:485    [-]
18   check 3.3.2.7 took 1.487671s
I220127 06:47:35.142591 1 workload/tpcc/tpcc.go:485    [-]
19   check 3.3.2.8 took 436.435ms
I220127 06:47:35.503616 1 workload/tpcc/tpcc.go:485    [-]
20   check 3.3.2.9 took 360.963ms
```

You can verify whether data for TPC-C is loaded or not by browsing the tpcc database and the tables, as shown in the following code snippet:

```
root@localhost:26257/defaultdb> show databases;
  database_name | owner | primary_region | regions |
survival_goal
  -------------+-------+----------------+---------+-----
-----------
    defaultdb  | root  | NULL           | {}      | NULL
    postgres   | root  | NULL           | {}      | NULL
    system     | node  | NULL           | {}      | NULL
```

```
     tpcc            | root  | NULL            | {}       | NULL
  (4 rows)
```

Here are the tables in the tpcc database:

```
root@localhost:26257/defaultdb> use tpcc;
SET


Time: 1ms total (execution 1ms / network 0ms)


root@localhost:26257/tpcc> show tables;
   schema_name | table_name | type  | owner | estimated_
row_count | locality
  --------------+------------+-------+-------+------------
  --------+-----------
     public      | customer   | table | root  |
  150000 | NULL
     public      | district   | table | root  |
  50 | NULL
     public      | history    | table | root  |
  150000 | NULL
     public      | item       | table | root  |
  100000 | NULL
     public      | new_order  | table | root  |
  45000 | NULL
     public      | order      | table | root  |
  150072 | NULL
     public      | order_line | table | root  |
  1500459 | NULL
     public      | stock      | table | root  |
  500000 | NULL
     public      | warehouse  | table | root  |
  5 | NULL
  (9 rows)
```

Once this is done, you can run a sample workload with the following command:

```
$ cockroach workload run tpcc --warehouses=5
--ramp=3m --duration=3m 'postgresql://root@
localhost:26257?sslmode=disable'
```

Here is a description of some of the options used in the preceding command:

- `--warehouse` is the number of warehouses that will be used for loading the data.

- `--ramp` is the duration for which the load will be ramped up, which will warm the cluster.

- `--duration` is the total duration to run the workload.

For both ramp and duration, time units can be specified in h, m, s, ms, us, and ns.

You will see statistics getting emitted every second on the console. Here is an example snapshot of statistics you might see:

_elapsed	errors	ops/sec(inst)	ops/sec(cum)	p50(ms)	p95(ms)	p99(ms)	pMax(ms)
145.0s	0	0.0	0.1	0.0	0.0	0.0	0.0 delivery
145.0s	0	0.0	0.4	0.0	0.0	0.0	0.0 newOrder
145.0s	0	0.0	0.1	0.0	0.0	0.0	0.0 orderStatus
145.0s	0	1.0	0.5	176.2	176.2	176.2	176.2 payment
145.0s	0	0.0	0.0	0.0	0.0	0.0	0.0 stockLevel
146.0s	0	0.0	0.1	0.0	0.0	0.0	0.0 delivery

Once the workloads are done running, you will see a performance metrics summary at the end. Here is a sample summary:

_elapsed	tpmC	efc	avg(ms)	p50(ms)	p90(ms)	p95(ms)	p99(ms)	pMax(ms)
180.0s	62.0	96.4%	179.5	176.2	243.3	260.0	352.3	469.8

This metrics summary includes an average of per-operation statistics. Similarly, you can run several other types of workloads that are provided out of the box. Please visit `https://www.cockroachlabs.com/docs/stable/cockroach-workload.html` to learn more.

In the next section, we will go over key aspects to consider before and during database migration.

Migration – Things to consider

In this section, we will go over some important factors that affect your migration strategy. Migrating between databases is always a tedious, error-prone, and complicated endeavor. It is also possible to get into nightmarish situations if you are not thorough and careful. Also, if your business is sensitive to downtime during this migration, you should find a mechanism for doing continuous migration. It is important to make sure you do your own benchmarking for the current schema, workload, and queries on CockroachDB before you decide to migrate your production data and workloads.

Here are some key aspects you should pay attention to before deciding on migration:

- **Cost**: You should be asking yourself whether the overall cost is going to go down or increase once you move to CockroachDB. Since the cost of maintaining a database can be significant, you should do a back-of-the-envelope calculation to see the cost difference.

- **Learning curve**: Although CockroachDB supports most SQL constructs, it's not 100% compatible with traditional databases. This requires modifying some existing queries and also doing benchmarking for those queries. If there are many features missing in CockroachDB that you are using in the current database, then all your application teams have to be educated about how to rewrite those queries.

- **Security**: When you are moving to CockroachDB, if the resources allocated for CockroachDB are within the same infrastructure as that of the previous database, then you will have fewer things to worry about in terms of security. For example, migrating to CockroachDB within your **on-premises cluster** is very different from migrating to **CockroachDB-as-a-Service** hosted by some third-party vendor. In the first case, all your security configurations can remain the same, whereas in the latter case, you will just get an endpoint to talk to and there is always a chance that the third-party infrastructure gets compromised and potentially exposes your data to bad players. So, it's always a good idea to run the migration plan through the **information security** (**InfoSec**) team and get their opinion.

- **Vendor lock-in**: Vendor lock-in happens when you are forced into using a specific product or a service, either because there are no other vendors who provide the same service or because you are already deeply invested in one. Vendor lock-in is a tricky subject, and it can happen in several ways. Here are some situations that can lead to vendor-lock-in:

A. Completely relying on an as-a-service offering by a third-party vendor

B. Extensively using all the enterprise-only features

C. Not having in-house expertise to deal with fires or to develop new features

D. Getting into long-term contracts with vendors because of an initial discount

I am sure there are plenty more points to discuss here, but the intention is to only highlight the important ones.

Once you have finalized that you are definitely going to migrate to CockroachDB, here are some key items you should plan for:

- **Capacity planning**: The number of nodes, memory, and storage space can all vary from what's allocated to a traditional database. So, make sure you provision enough resources upfront in order to avoid any issues arising from a lack of resources.

- **Downtime**: Although you will have tried the migration several times in a pre-production environment before proceeding to production, you can always encounter new issues if the scale and data of production and pre-production are different. So, just be prepared for all sorts of failures, keep the stakeholders informed, and line up the necessary human resources before you start the migration.

- **Migrating data in smaller chunks**: Instead of dumping the entire database at once, you can just migrate one table at a time. This can finish relatively faster compared to the entire database import, and you can also avoid timeouts.

- **Maintaining data in source and target databases for some time**: It would be advisable to maintain the data in both source and target databases for several months till you are comfortable with CockroachDB. Maintaining the data in the source database will allow you to switch back to the source database if things don't go as expected after the migration. But this definitely creates additional complexity, as you now have to maintain the data in two different databases and make sure the data is still consistent between them.

In the next section, we will learn how to migrate from traditional databases into CockroachDB.

Migrating from traditional databases

In this section, we will outline a few generic steps that are involved in migrating data between any two databases. After that, we will go over a specific example of migrating data from PostgreSQL into CockroachDB.

Migrating from other databases into CockroachDB usually involves the following generic steps:

1. Export the schema from the source database.

2. Export the data from the source database.

3. Transform the data into a desirable format. In most cases, **comma-separated values (CSV)** should work just fine.

4. Compress and transfer the schema and the data to the desired location where it can be imported into CockroachDB.

5. Do the data type mapping in the schema from the source to the CockroachDB database.

6. Import the schema, along with the data, into CockroachDB.

7. Manually create users and privileges.

8. All the application teams also have to map current queries from the source database to CockroachDB.

Now, let's look at a specific example of how this migration looks when we migrate from PostgreSQL into CockroachDB.

Migrating from PostgreSQL to CockroachDB

Here are the steps to take to migrate from PostgreSQL:

1. Dump the database: pg_dump is a utility used for taking backups for a PostgreSQL database. You can use the following command to dump a given database:

   ```
   pg_dump my_database > my_database.sql
   ```

2. Map the required data types in the schema. For example, any new types of data that are created in PostgreSQL using CREATE TYPE will not be supported in CockroachDB as they are not the standard data type. So, these should be manually mapped to CockroachDB data types in the previously dumped file.

3. Compress and move the dump to a place where it can be imported into a CockroachDB cluster.

4. Import the schema and data using the following command:

   ```
   IMPORT PGDUMP 'nodelocal:///my_database.sql.gz' WITH
   ignore_unsupported_statements;
   ```

5. Once this is done, you can list the databases and tables and run sample queries on the imported database to make sure it's imported correctly.

6. The final step is to verify that the entire data is imported without any missing data. For this, you can compare all the tables and row counts of the imported tables with the source database. For critical tables, you can dump it from the source and target database and compare the byte size and contents.

Data migration is always tedious and can fail for various reasons. So, instead of making it a one-time activity, you can also explore the option of continuous migration and keeping the previous database in tandem with CockroachDB. This is required till you are completely confident that CockroachDB is performing well for all your use cases and all the data is migrated to CockroachDB cluster. For continuous migration, you can publish the change data capture from a traditional database into Kafka and consume the events to make relevant updates in CockroachDB.

Summary

In this chapter, we discussed aspects that contribute to performance. We also looked at popular benchmark suites and went over the process of benchmarking a CockroachDB cluster. Next, we learned about important things we should consider before and during migration. Lastly, we familiarized ourselves with generic steps for migrating from some other database to CockroachDB and discussed a specific example of migrating from PostgreSQL to CockroachDB.

With that, we have come to the end of this book. All the code used in this book has been shared at `https://github.com/PacktPublishing/Getting-Started-with-CockroachDB`. Please make sure you are using the latest version of CockroachDB when trying out these examples. All the examples have been tried with v21.2.0. Please reach out to the author if you find any mistakes. Thank you for reading this book— I hope you enjoyed the contents of this book and are now ready to start exploring CockroachDB as a potential database for your use cases.

Appendix: Bibliography and Additional Resources

Here is a list of references used in the book:

- Brian W. Kernighan and Alan A. A. Donovan: *"The Go Programming Language"*, 2015

- Eric Brewer: *"CAP twelve years later: How the 'rules' have changed"*, Computer, Volume 45, Issue 2, pg. 23–29, 2012

- Ongaro, Diego and John K. Ousterhout: *"In Search of an Understandable Consensus Algorithm."* USENIX Annual Technical Conference, 2014

- Ramez Elmasri and Shamkant B. Navathe: *"Fundamentals of Database Systems"*, 7th edition, 2016

- Rebecca Taft et al.: *"CockroachDB: The Resilient Geo-Distributed SQL Database."* In Proceedings of the 2020 ACM SIGMOD International Conference on Management of Data (SIGMOD'20), Portland, OR, USA, June 2020

- Thomas H. Cormen, Charles E. Leiserson, Ronald L. Rivest, and Clifford Stein: *"Introduction to Algorithms"*, 3rd edition. MIT Press, ISBN 978-0-262-03384-8, pp. I-XIX, 1-1292, 2009

Here is a list of additional resources that you can refer to:

- `https://www.cockroachlabs.com`
- `https://www.cockroachlabs.com/docs/v21.2/`
- `https://github.com/cockroachdb/cockroach`
- `https://s2geometry.io/`
- `https://www.royans.net/2010/02/brewers-cap-theorem-on-distributed.html`
- `http://blog.thislongrun.com/2015/03/the-confusing-cap-and-acid-wording.html`

Index

A

abstract syntax tree (AST) 25
ACID properties
  about 48
  atomicity 48
  consistency 49
  durability 51
  isolation 49, 50
advanced debugging options
  about 196
  advanced debugging 197
  configuration 197
  reports 196
Advanced Encryption Standard (AES) 177
ALTER INDEX
  parameters 129
ALTER statement
  about 129
  example 130
ALTER TABLE
  parameters 129
Amazon Elastic Block Store (EBS) 86
Amazon Elastic Compute Cloud (EC2) 86
Amazon Web Services 66
American National Standards
    Institute (ANSI) 24

Apache Derby 12
arrays 110
atomicity 48
atomicity, consistency, isolation,
    and durability (ACID) 31
atomic transactions
  without parallel commits 52, 54
  with parallel commits 55, 56
audit logging 168-181
authentication 169
authorization 168
authorization mechanisms
  about 174, 175
  privileges 176, 177
  roles 175, 176
automatic rebalancing 96-98
availability and partition
    tolerance (AP) 14
availability zones 64, 65

B

Berkeley DB 12
Bigtable 16
bridge network 21

C

C++ 12
California Consumer Privacy
    Act of 2018 (CCPA) 63
capacity planning 192
CAP-available system
  about 14
  example 15
CAP-consistent system
  about 13
  example 14
CAP theorem (Consistency, Availability,
    and Partition Tolerance)
  about 13, 49
  availability and partition
    tolerance (AP) 14
  consistency and availability (CA) 15
  consistency and partition
    tolerance (CP) 13
cardinality
  about 6
  many-to-many relationship 7
  one-to-many relationship 6, 7
  one-to-one relationship 6
Cassandra 5, 15
central processing unit (CPU) 200
certificate authority (CA) 170
certificates and keys
  generating 171, 172
change data capture (CDC) 12, 157
ciphertext 177
cleanup.sh script
  reference link 99
client authentication
  about 169, 170, 172
    Generic Security Services API (GSSAPI),
      with Kerberos authentication 173

password authentication,
    without TLS 172
password authentication, with TLS 173
single sign-on (SSO) authentication 173
cloud 64
cloud computing 5
cloud provider 64, 201
cloud regions 64
cluster upgrade
  issues, avoiding 193, 194
COBOL language 4
CockroachDB
  about 5, 15
  atomicity 51, 52
  consistency 57
  databases 125
  durability 57-60
  functional layers 18
  high-level overview 17
  inspiration 16
  interacting, with disk 41
  isolation 57
  key concepts and terms 16
  layers 19
  migrating, from PostgreSQL 209, 210
  migrating, from traditional
    databases 208, 209
  need for 16
  performance benchmarking 204-206
  used, for achieving geo-partitioning 67
CockroachDB admin user
    interface 146-148
CockroachDB-as-a-Service 207
CockroachDB cluster
  overview 148-151
CockroachDB cluster setup
  reference link 189

CockroachDB configurations
  node configuration 192
  operating system 192
  storage 192
CockroachDB features
  rebalancing 68
  replication 68
  resiliency 68
CockroachDB, log channel
  DEV 188
  HEALTH 188
  OPS 188
  PRIVILEGES 188
  SENSITIVE_ACCESS 188
  SESSIONS 188
  SQL_EXEC 188
  SQL_INTERNAL_PERF 188
  SQL_PERF 188
  SQL_SCHEMA 188
  STORAGE 188
  TELEMETRY 188
  USER_ADMIN 188
CockroachDB, log levels
  ERROR 187
  FATAL 187
  INFO 187
  WARNING 187
CockroachDB, technical advisories
  reference link 189
CockroachDB, topology patterns
  multi-region 70, 71
  single region 67
Cockroach debug logs
  obtaining 189, 190
cockroach debug zip command
  options 190

CODASYL (Conference/Committee
    on Data Systems Languages) 4
column-level constraints
  about 137
  CHECK <condition> 137
  DEFAULT constraint 138
  FOREIGN KEY 138, 139
  NOT NULL 139
  PRIMARY KEY 139
  UNIQUE constraint 139
comma-separated values (CSV) 209
common table expressions (CTEs) 131
Compute Engine 87
configuration-related issues
  about 193
  clock synchronization error 193
  open file descriptor limit 193
  port, using 193
  storage directory exist 193
connection issues 190
connection refused 190
consensus 34
consistency 49
consistency and availability (CA) 15
consistency and partition
    tolerance (CP) 13, 14, 57
Continuous Integration/Continuous
    Deployment (CI/CD) 67
Coordinated Universal Time (UTC) 137
Couchbase 5
Couchbase Lite 12
CouchDB 15
CREATE DATABASE statement
  parameters 124
create, read, update, and delete
    (CRUD) 12, 42
CREATE statement 124

CREATE TABLE syntax
  about 125
  parameters 125, 126
  reference link 126
CREATE VIEW
  about 126
  example 127, 128
  parameters 127
C Yacc 25

D

data at rest 168, 177
database
  about 3
  evolution 4
  history 4
  NewSQL 5
  NoSQL 5
  object-oriented databases 4
  SQL 4
database concepts
  about 6
  cardinality 6
  database models 8
  database storage engine 12
  embedded databases 12
  mobile database 12
  processing models 11
database index 102
database management system
    (DBMS) 3, 42, 51
database migration
  considerations 207, 208
database models
  about 8
  hierarchical database model 8
  network model 9

object-relational model 10
relational model 9
Databases dashboard 158-160
database storage engine 12
data definition language (DDL)
  about 124, 154
  ALTER statement 129
  CREATE statement 124
data distribution
  across multiple nodes 32
Datadog 186
data encryption at rest 177, 178
data encryption in flight 177, 178
data keys 177
data manipulation language (DML) 131
data query language (DQL)
  about 135
  SELECT statement 135
data replication
  for resilience and availability 34
data types 136
DB2 Everyplace 12
debug logs
  collecting 186
decryption 177
default connections
  database 125
default database
  demonstration purposes 125
DELETE statement
  about 134
  example 134
  parameters 134
distribution layer 18
Docker
  used, for installing single-node
    CockroachDB cluster 21-23
document model 10

Dropbox 5
DROP statement
  about 130
  example 130
DROP TABLE
  parameters 131
duplicate indexes 80, 81
durability 51
Dynamo 5

E

Edgestore 5
election timeout 35
embedded databases 12
encryption 177
entity-relational model 10
external sink
  logs, sending to 189
Extract, Transform, and Load (ETL) 11

F

Facebook 5
fault domains 67
fault tolerance
  achieving 86
  achieving, at storage layer 86
  working example 87-96
follower reads 83
follow-the-workload 83, 84
FORTRAN language 4
FoundationDB 5
FULL JOIN 141
full table scan 102
functional layers, CockroachDB
  distribution layer 18
  replication layer 18

SQL layer 18
storage layer 18
transactional layer 18

G

General Data Protection
      Regulation (GDPR) 64
geo-partitioned leaseholders 75-79
geo-partitioned replicas 71-75
geo-partitioning
  about 62, 63
  achieving, with CockroachDB 67
  advantages 64
gibibytes (GiB) 200
GitHub 16
Golang 12, 43
Golang ORM (GORM) 24
Google 5
Google Cloud Platform (GCP) 66, 87
Google Spanner 5
Goyacc 25
GreenPlum 5
GSSAPI setup
  reference link 173

H

hash-sharded indexes 109, 110
Hbase 5
hierarchical database model
  about 8
  example 8
high-level overview 17
hot ranges 156
HSQLDB 12
hybrid cloud 64

hybrid transaction/analytical
     processing (HTAP) 12
HyperText Transfer Protocol
     (HTTP) requests 22

I

IBM DB2 4, 8
IDMS (Integrated Database
     Management Systems) 9
indexes
   about 102
   best practices 118, 119
   duplicate indexes 110, 111
   hash-sharded indexes 108-110
   inverted indexes 110
   partial indexes 111-113
   primary indexes 104-106
   secondary indexes 106, 107
   spatial indexes 113-115
   table joins 115-118
   types 103
   working 102, 103
InfinityDB 12
Information Management
     System (IMS) 4, 8
information security (InfoSec) 207
infrastructure
   configuring, considerations 200, 201
INNER JOIN 140
INSERT statement
   parameters 131
Integrated Data Store (IDS) 4, 9
Interactive Graphics Retrieval
     System (INGRES) 4
Internet Protocol version 4 (IPv4) 136
Internet Protocol version 6 (IPv6) 136

inverted indexes 110, 111
IP allowlisting 182
isolation 49
isolation levels
   read committed 50
   read uncommitted 49
   repeatable read 50
   serializable 49
   snapshot 49

J

Jepsen 57
Jobs dashboard
   jobs, tracking 163-165
JOIN types
   about 140
   FULL JOIN 141
   INNER JOIN 140
   LEFT OUTER JOIN 140
   RIGHT OUTER JOIN 141
JSON Web Token (JWT) 173

K

key concepts and terms, CockroachDB
   cluster 16
   leader 16
   leaseholder 16
   node 16
   raft log 16
   ranges 16
keys 102
known limitations
   reference link 143

L

leader election 35
leaseholder 26
LEFT OUTER JOIN 140
LevelDB 12
lexical analysis 24
log channel 188
log files 186, 187
logical plan 25
logical processors 30
log levels 187
log replication 36-41
logs
    sending, to an external sink 189

M

many-to-many relationship 7
MariaDB 4
memory 200
meta ranges 32, 33
metrics
    about 151
    categories 151-158
Microsoft Azure 66
mobile databases 12
MongoDB 5
monolithic sorted key-value
        store (MSKVS)
    about 32
    system data 32
    user data 32
multi-cloud 64
multi-node failures
    recovering from 98, 99
multi-region
    about 70, 71

duplicate indexes 80-82
follower reads 83
follow-the-workload 83, 84
geo-partitioned leaseholders 75-79
geo-partitioned replicas 71-75
Multiversion concurrency
        control (MVCC) 31, 52
MySQL 4, 8

N

navigational databases 4
Neo4j 5
network 201
network latency 194, 195
network model
    about 9
    example 9
network security 182
network time protocol (NTP) 193
NewSQL 5
node authentication 169-173
non-vectorized query execution 30
NoSQL 5

O

object model 10
object-oriented database systems
        (OODBMSes) 4
object-relational mappers (ORMs) 24
object-relational model 10
objects 110
one-to-many relationship 6, 7
one-to-one relationship 6
online analytical processing (OLAP) 11
online event processing (OLEP) 11

online transaction processing
    (OLTP) 11, 201
on-premises cluster 207
OpenSSL tool
    about 172
    download link 172
operating system 200
Oracle 8
Oracle Cloud 67
Oracle database Lite 12

P

parallel commits 51
partial index 111-113
Pebble 12, 42
performance, considerations
    about 200
    infrastructure 200, 201
    popular benchmark suites 201, 202
    specific use cases,
        benchmarking 202, 203
personally identifiable
    information (PII) 189
phantom read 50
point-in-time restore 181
PostgreSQL
    migrating, to CockroachDB 209, 210
Post Ingres (Postgres) 4, 24
primary indexes 104-106
primary key (PK) 104, 129
private cloud 64
processing models
    about 11
    hybrid transaction/analytical
        processing (HTAP) 12
    online analytical processing (OLAP) 11
    online event processing (OLEP) 11

online transaction processing
    (OLTP) 11
public cloud 64
public-private keys 169

Q

quadtree data structure 113
QUEL 4

R

Raft 34, 54
raft log 16
Raft nodes
    candidate 35
    follower 35
    leader 35
Raima Database Manager (RDM) 9
random-access memory (RAM) 152, 200
ranges 32
real-world benchmarking 202
recovery point objective
    (RPO) 168, 181, 182
recovery time objective
    (RTO) 168, 181, 182
Red Hat Enterprise Linux (RHEL) 200
region 65
region and zone, cloud providers
    Amazon Web Services 66
    defining 66
    Google Cloud Platform 66
    Microsoft Azure 66
    Oracle Cloud 67
region, cloud providers
    cloud cost 65
    data compliance 65
    latency 65

multi-cloud and hybrid
    cloud strategy 65
  services and features 65
relational model
  about 9
  example 10
remote procedure call (RPC) requests 35
replication factor 34, 86
replication layer 18
resharding 200
Riak 15
RIGHT OUTER JOIN 141
RocksDB 12

S

S2 geometry library
  reference link 113
schema 126
schema changes
  benefits 143
  limitations 143
  managing 142
secondary index 106, 107
security, best practices
  about 183
  automation 183
  backups and archived data 183
  certification and key rotation 183
  client password 183
  data encryption 183
  disaster recovery, planning 183
  personally identifiable
      information (PII) 183
  production access 183
  secret management 183
  Transport Layer Security (TLS) 183

security concepts 168, 169
SELECT statement
  about 135
  parameters 135
sequences
  example 141, 142
  using 141, 142
serializable snapshot 57
service-level agreement (SLA) 203
Sessions dashboard 160, 161
SHOW SQL syntax 136
single-node CockroachDB cluster
  installing, with Docker 21-23
single region
  about 67
  basic production 68, 69
  development 67
single sign-on (SSO) 173
sink
  about 189
  parameters 189
Site Reliability Engineering
    (SRE) 182, 188
slow queries
  tracking 191
Snowflake Schema 10
solid-state drives (SSDs) 201
Sorted Strings Tables (SSTable) 154
Spanner. See Google Spanner
spatial indexes 113-115
SQL Anywhere 12
SQLBoiler 24
SQLite 12
SQL layer
  about 18
  logical planning 25
  parsing 25
  physical planning 26, 27

SQL parser 24
SQL query execution 24, 30
SQL Server Compact 12
SQL Server Express 12
SQL statements 24
Star Schema 10
storage engine 42
storage input/output (I/O) 201
storage layer 18
store keys 177
Structured English Query
    Language (SEQUEL) 4
Structured Query Language (SQL) 4, 201
swap memory 152
Sybase 4
synthetic benchmarking 202
sysbench 202

T

table 158-160
table data 34
table joins
    about 140
    indexes 115-118
TAO 5
tebibytes (TiB) 201
TPC-H 201
traditional databases
    migrating, to CockroachDB 208, 209
transaction
    phases 31
transactional key-value store
    managing 31
transactional layer 18, 31
transaction identifier (ID) 31

transaction log 51
Transaction Processing Performance
    Council Benchmark C (TPC-C) 201
transaction records 52
Transactions dashboard 162, 163
transaction status recovery 56
Transport Layer Security (TLS) 173

U

Uniform Resource Locator (URL) 27
Universally Unique Identifier (UUID) 71
UPDATE statement
    example 132
    parameters 132
UPSERT statement
    about 133
    parameters 133

V

Vault
    reference link 174
vectorized query execution 30
view 126
virtual CPUs (vCPUs) 200
virtual private cloud (VPC) 182
Vitess 5
volumes 21
VPC peering
    advantages 182

W

write intent 49, 52

Y

Yahoo! Cloud Serving Benchmark
      (YCSB) 89, 202
Yet Another Compiler-
      Compiler (Yacc) 24
Youtube 5
YugabyteDB 5

Z

zone 65

Packt.com

Subscribe to our online digital library for full access to over 7,000 books and videos, as well as industry leading tools to help you plan your personal development and advance your career. For more information, please visit our website.

Why subscribe?

- Spend less time learning and more time coding with practical eBooks and Videos from over 4,000 industry professionals

- Improve your learning with Skill Plans built especially for you

- Get a free eBook or video every month

- Fully searchable for easy access to vital information

- Copy and paste, print, and bookmark content

Did you know that Packt offers eBook versions of every book published, with PDF and ePub files available? You can upgrade to the eBook version at packt.com and as a print book customer, you are entitled to a discount on the eBook copy. Get in touch with us at customercare@packtpub.com for more details.

At www.packt.com, you can also read a collection of free technical articles, sign up for a range of free newsletters, and receive exclusive discounts and offers on Packt books and eBooks.

Other Books You May Enjoy

If you enjoyed this book, you may be interested in these other books by Packt:

Mastering PostgreSQL 13 - Fourth Edition

Hans-Jürgen Schönig

ISBN: 9781800567498

- Get well versed with advanced SQL functions in PostgreSQL 13
- Get to grips with administrative tasks such as log file management and monitoring
- Work with stored procedures and manage backup and recovery
- Employ replication and failover techniques to reduce data loss
- Perform database migration from Oracle to PostgreSQL with ease
- Replicate PostgreSQL database systems to create backups and scale your database
- Manage and improve server security to protect your data
- Troubleshoot your PostgreSQL instance to find solutions to common and not-so-common problems

Packt is searching for authors like you

If you're interested in becoming an author for Packt, please visit authors. packtpub.com and apply today. We have worked with thousands of developers and tech professionals, just like you, to help them share their insight with the global tech community. You can make a general application, apply for a specific hot topic that we are recruiting an author for, or submit your own idea.

Share Your Thoughts

Now you've finished *Getting Started with CockroachDB*, we'd love to hear your thoughts! Scan the QR code below to go straight to the Amazon review page for this book and share your feedback or leave a review on the site that you purchased it from.

https://packt.link/r/1-800-56065-6

Your review is important to us and the tech community and will help us make sure we're delivering excellent quality content.